St. Thomas Aquinas
Siger of Brabant
St. Bonaventure

On the Eternity of the World

(De Aeternitate Mundi)

Second Edition

Translated from the Latin
With Introductions

By

CYRIL VOLLERT, S.J., S.T.D.
Professor of Theology
St. Mary's College

LOTTIE H. KENDZIERSKI, Ph.D.
Professor of Philosophy
Marquette University

PAUL M. BYRNE, L.S.M., Ph.D.
Assistant Professor of Philosophy
Marquette University

Marquette University   Milwaukee, WI   U.S.A.

Library of Congress Card Number: 64-7796
Copyright, 1964, 1984, The Marquette University Press
Milwaukee, Wisconsin
Manufactured in the United States of America
I.S.B.N. 0-87462-216-6

CONTENTS

ST. THOMAS AQUINAS

SIGER OF BRABANT

ST. BONAVENTURE

Preface

This volume makes available in English translation the works of St. Thomas Aquinas and Siger of Brabant entitled *On the Eternity of the World* as well as a few texts on the topic from various works of St. Bonaventure. If the order of presentation were chronological, which it is not, it would probably be St. Bonaventure, St. Thomas, and then the work of Siger. Since, however, the translation of St. Thomas' *De Aeternitate Mundi contra murmurantes* as well as other texts on this subject by Fr. C. Vollert, S.J., is preceded by an excellent and extensive introduction to the whole problem, we have decided to lead off with this work. This is followed by L. H. Kendzierski's introduction to and translation of Siger of Brabant's *De Aeternitate Mundi*. Finally, there is appended a translation of some texts from various works of St. Bonaventure, inevitably lifted out of a complexly interwoven context, together with a brief introduction by P. M. Byrne.

Why present twentieth century readers with thirteenth century views on the eternity, or more precisely perpetuity of the world? Very simply, the mysteries and problems, questions and intelligibilities involved in and implied by our speaking of an *eternal* God *creating* a temporally *limited* universe have always been "contemporary" ever since men in the Judeo-Christian intellectual tradition have reflected upon their *linear* view of time and history as over against the *cyclic* conception of the pagan Greek. Evidence for this "contemporary" character ranges back in time from current periodical literature to the *Acts of the Apostles*. The 1964 Presidential Address to the Metaphysical Society of America is entitled "Toward a Metaphysics of Creation,"[1] but the first confrontation of "the metaphysic of the Bible and the metaphysic of the Gentiles" is described for us in Acts 17: 16-33 . There we read of that wonderful scene at the Areopagus in Athens when St. Paul brought something new to the Greek philosophers, namely, the absolute beginning to be of a creature totally dependent for its being on a Creator or, in other words, the very "newness" of the world itself. "No townsman of Athens, or stranger visiting it, has time for anything else than saying something new, or hearing it said." This interesting human comment on Athenian ways appears within the account that follows.

---

1. Peter A. Bertocci, "Toward a Metaphysics of Creation," in *Review of Metaphysics* , XVII (June, 1964), 493-510.

And while Paul was waiting for them in Athens, his heart was moved within him to find the city so much given over to idolatry, and he reasoned, not only in the synagogue with Jews and worshippers of the true God, but in the market place, with all he met.

He encountered philosophers, Stoics and Epicureans, some of whom asked, "What can his drift be, this dabbler?" While others said, "He would appear to be proclaiming strange gods"; because he had preached to them about Jesus and the Resurrection. So they took him by the sleeve and led him up to the Areopagus: "May we ask," they said, "what this new teaching is thou art delivering? Thou dost introduce terms which are strange to our ears; pray let us know what may be the meaning of it."

So Paul stood up in the full view of the Aeropagus, and said, "Men of Athens, wherever I look I find you scrupulously religious. Why, in examining your monuments as I passed by them, I found among others an alter which bore the inscription, *to the unknown God* . And it is this unknown object of your devotion that I am revealing to you.

"The God who made the world and all that is in it, that God who is the Lord of heaven and earth, does not dwell in temples that our hands have made; no human handicraft can do him service, as if he stood in need of anything, he, who gives to all of us life and breath and all we have. It is he who has made, of one single stock, all the nations that were to dwell over the whole face of the earth. And he has given to each the cycles it was to pass through and the fixed limitations of its habitation, leaving him to search for God; would they somehow grope their way towards him? Would they find him? And yet, after all, he is not far from any one of us; it is in him that we live, and move, and have our being; thus, some of your poets have told us, for indeed we are his children.[2]

This familiar passage has been the subject of the following incisive comment by Claude Tresmontant:

---

2. Acts 17:16-29

By declaring, in the very heart of Athens, that God created the cosmos, St. Paul made a frontal attack on the fundamental principle of all the philosophy of antiquity. According to that philosophy, the cosmos is God, uncreated, existing from eternity; it has no need of a creator, it is all-sufficient, necessary, it is consistency itself. At most, it requires a demiurge to put it in order, for order is preceded by chaos. For Aristotle, the stars are gods, "distinct substances," eternal, outside any becoming; astronomy is not a physical science but a theology. Nor can the uncreated stars ever perish. Since a becoming, from birth to death, must be recognized in our sublunary world, it was said to be cyclic, recurring: time chases its own tail. This is the "endless returning" of the metaphysics, cosmogonies and mythologies of pagan antiquity.[3]

So it was when St. Paul brought something new in theology and philosophy to the philosophers of Athens in the first century.

The situation is almost exactly the reverse when we turn to thirteenth century Paris. Now the theologian-philosophers of Paris were being confronted with something "new," the moving cause, the eternal motion and eternally moved world of Aristotle's *Physics* and Averroes' commentaries as well as the necessarily, and so eternally, produced world in the Avicennian version of Aristotle's thought. The suspicion of and hostile reaction to this new naturalistic Greek world-view leading to the condemnations of 1210, 1215, and 1270 is of course well known. St. Bonaventure led what Gilson had aptly termed the "theological reaction" that eventually culminated in the condemnation of 1277.[4] In the title of St. Thomas Aquinas' work *De Aeternitate Mundi contra murmurantes*, the "murmurantes" refers to the Augustinian theologians, and in particular St. Bonaventure, of whom St. Thomas is quite critical and whose arguments for a temporal creation he regards as entirely ineffective, if not ludicruous. St. Thomas will admit the rational possibility of an eternal world, but he will not grant, nor will his own metaphysics of efficient creative causality permit him to grant, any compelling force to the arguments of Aristotle the *Physicus* for the eternity of time and motion. Siger of Brabant, on the other hand, while not denying the absolute beginning of creatures that faith teaches, is philosophizing as a Master in Arts quite

3. Claude Tresmontant, *Saint Paul and the Mystery of Christ*, trans. by Donald Attwater, (New York: Harper Men of Wisdom, 1957), p. 132.
4. E. Gilson, *History of Christian Philosophy in the Middle Ages* (New York: Random House, 1955), pp. 402ff.

unconcerned with, if not indifferent to, the difficulties Aristotle's texts as commented by Averroes were causing the theologians. Speaking only as a philosopher, he thinks that he can accordingly adduce what might be called the Aristotelian *rationes necessariae* for the eternity of the world. Thus the thirteenth century controversy on this topic arose in response to a novelty, namely, the "new" works of Aristotle and his commentators.

It would be a mistake, however, to regard this controversy as being merely a matter of how to deal with the "formidable" Aristotle. Philosophical questions with a profound importance both in themselves as well as for the later history of philosophy were inextricably interwoven into the discussions. One such question that might be kept in mind when reading the texts is that of the very possibility of "new" Christian philosophies.

St. Augustine used his great rhetorical talent to highlight the truths known to the faith-illumined and reflective intellect. By dialectical analysis, by metaphors and images, by word analysis, by historical allusions, by exposing the interior depths of his own being, he shared with others that understanding which was the reward of his own faith. In this way he developed the *philosophia nostra*, the Christian wisdom that results from a faith driven by love to seek understanding. But he does not speak of *rationes necessariae*, i.e., of demonstrations that are proofs compelling the assent of the intellect of the other. It is, of course, St. Anselm who is famous for presenting Augustinian wisdom in the rational mode of the *rationes necessariae*. Reason flows from a faith-illumined understanding and in that situation its highest function is to work out the arguments showing the necessity involved in any of the truths already known to that faith-illumined intellect. By the time St. Thomas Aquinas appears on the scene in the thirteenth century, this Anselmian attitude toward faith, understanding, and reason seems to have become entrenched as the Christian mode of philosophizing, if not the only properly Christian philosophy. This attitude had its consequences, of course, for the controversy that arose over the eternity of the world.

St. Bonaventure not only held that we know by faith and revelation that the world, motion, and time all had a beginning, he also asserted that reason can produce many arguments showing that such *must* be the case. St. Thomas regarded the beginning of the world and time as a non-necessitated event or fact (in the sense of a "something freely done"). It is not a necessary truth to which we can reason, such as the proposition that God is, but rather it is knowable as true only to Him in whose presence it occurs and to others only inasmuch as He "reports", that is, reveals it. Thus, by revelation we know it to be true, but reason cannot demonstrate either that the world had to have a beginning (St. Bonaventure) or that it could not (Aristotle and Siger of Brabant). In general, even as the so called Latin Averroists were proclaiming philosophy's absolute autonomy and the separation of faith and reason, St. Thomas was proposing another, a new Christian philosophy which *could* integrate within itself by way of reason operative in its own *proper* sphere many of the basic elements of Aristotle's teaching. In the eyes of his many traditionalist-minded opponents this proposal was his unforgivable sin, *his* "novelty." What his contemporaries called the "novelties of Thomas" sprang from his conception of the way in which the man of Christian faith should philosophize. For his position implied that a quite new Christian philosophy was not only possible, but absolutely necessary in the face of the intellectual challenge raised by the new Aristotelianism. The condemnation of 1277, in part at least, was a disastrously non-philosophical resolution to the philosophical question of whether a new Christian philosophy was needed. The philosophical work of St. Thomas Aquinas had raised that fundamental question, among many others. It would appear that very few understood the question he was raising or appreciated his answer.

August 7, 1964   P.M.B.

Preface to the Second Edition

This second edition is dedicated to the memory of Dr. Edward D. Simmons, Vice-President for Academic Affairs at Marquette University, whose support made its preparation possible.

The second edition of *On the Eternity of the World* has incorporated only minor corrections into the text of the first edition. For the second edition the text has been completely reformatted and typeset anew with the result that pagination here differs slightly from the first edition.

This edition was produced on a Hewlett-Packard HP+ Laser Printer with a Tandy Model 4 microcomputer using the following production software [registration and copyright owners in braces]: LS-DOS operating system [Logical Systems, Inc., Grand Junction, CO], with Allwrite editor and formatter [PROSOFT, Inc., North Hollywood, CA], Hypercross disk transfer utility [Hypersoft, Raleigh, NC], and a variety of programming and utility software [MISOSYS, Sterling, VA].

The computer typesetting and laser printing of this text was carried out by Ms. Janette Hodge and myself with the invaluable technical assistance of Dr. Lee C. Rice. Assistance with the typing of the entire manuscript in ASCII format by Mrs. Kathleen A. Hawkins is very gratefully acknowledged.

Richard C. Taylor
Chair, MPTT Editorial Board
December 8, 1987

St. Thomas Aquinas

On the Eternity
of the World

(De Aeternitate Mundi)

And Selected Miscellaneous Texts

Translated from the Latin
With an Introduction

By

CYRIL VOLLERT, S.J., S.T.D.
Professor of Theology
St. Mary's College

Translator's Introduction

Thinking men have always been interested in the question of the age of the world. When did the vast history of the universe begin? If we look forward toward the future, we are not aware of any reason that compels us to assert that the material component of the world must have an end. If we roam back in thought toward the past, can we not conceive a universe that never began? Or is a beginning imperative? Scientists, philosophers, and theologians have brought the resources and procedures of their respective disciplines to bear on such questions. The answers they offer not only fluctuate between opinion and confidence of certitude, but vary from affirmative to negative.

I. Duration of the Universe According to Science and Theology

By examining the structure, movements, and laws of the cosmos, the sciences of astronomy, cosmology, and astrophysics have been able to push back into the past until they arrived at a form of the universe which is the origin of the form it exhibits today. They have also been able to calculate the duration of its history. A striking array of proofs, based on studies of radioactive elements, the earth and our solar system, meteorites, double stars and star-clusters, the Milky Way galaxy, and the billions of other galaxies within range of optical and radio telescopes, points to an age of some five or six billion years for the universe.[1] The very convergence of dates arrived at by diverse methods applied to diverse sources of evidence intimates that an order of magnitude of this sort is free from serious error. All the procedures employed to measure the age of the universe flow together toward an initial hour of time. The researches undertaken by astrophysics and cosmologists lead them back to a moment when there were no elements, no plants, no sun, no stars, no galaxies. That moment, the farthest frontier reached by science in its retrogressive tracing of the world's evolution, defines the start of the cosmos.

----

1. Cf. C. Vollert, S.J., "Origin and Age of the Universe Appraised by Science," *Theological Studies,* XVIII (1957), 137-68.

Does this zero hour mark the origin of matter by creation from nothing, or does it signalize only the inception of the present universe? Is the beginning absolute or relative? Opinions are so sharply divided that an attitude of reserve seems to be indicated. Science cannot answer all questions, and many scientists are aware that the question about what may have preceded the origin of the universe as now known exceeds its competence. At any rate, science has not definitely demonstrated the first day of the world. Something behind the most primitive state to which science has penetrated is at least conceivable, for matter may have endured indefinitely in a condition that simply eludes us. Theories pointing to a finite past may well be regarded as highly probable, but no scientific discoveries have eliminated the opposite alternative.

Indeed, the question whether the world had a beginning or not, does not fall within the domain of science. The function of science is to retrace the chain of events in the past by going back toward the origins of the universe. However, since it abides in this order of relations between successive steps, science as such is not in a position to decide whether, prior to a certain state, further time is admissible or excluded. Consequently, solution of the problem, whether the world has a finite or an infinite duration, belongs to another sphere of investigation.

Certainly the sources of faith teach that there was a beginning of time. The question of fact is revealed and has been defined by the Church. The first verse of Genesis mentions the origin of the universe, though without metaphysical precision. Elsewhere the Bible teaches that prior to the world, before the earth and the universe was formed, before anything whatever was created, God exists eternally and possesses His wisdom by which He made all things (Psalm 89:2; Proverbs 8:22). Jesus Christ speaks of the glory which He had with the Father "before the world was" (John 17:5); in Christ, the Father has chosen His elect "before the foundation of the world" (Ephesians 1:4). These and similar texts indicate the eternity of the divine life that transcends the limited duration of the universe.

On the basis of revelation, as conveyed in the Scriptures and an undeviating tradition, the Church has repeatedly taught, in the face of error, that God alone lacks beginning and is eternal, and that He created the universe in a condition of successive duration following on a first moment. An occasion for such a declaration occurred in the twelfth century, when the Albigensian heresy renewed Manichaean dualism; this doctrine proposed two eternal principles, one good, from which spirits proceed, the other evil, from which matter issued. Against this error the Fourth Council of the Lateran defined, in 1215, that God alone has no beginning but always is and always will be; the eternal God is the one and only principle of all things, "Creator of all things visible and invisible, spiritual and corporeal; by His almighty power, at the beginning of time

He created both orders of creation alike out of nothing, the spiritual and the corporeal world, the angelic and the material."[2] The First Vatican Council proclaims the same truth against various forms of pantheism: "The one and only true God . . . by a completely free decision, at the beginning of time created both orders of creatures alike out of nothing, the spiritual and the corporeal."[3]

What is the value of these doctrinal decrees? Some doubt has been raised whether the two Councils intended to define the particular circumstance "at the beginning of time."[4] In both Councils the phrase seems to be mentioned as though in passing. The Lateran Council unquestionably teaches against the Albigensians that all creatures, material as well as spiritual, were created by God. It adds that angels are incorporeal and that creation dates from the beginning of time; but these clarifications seem to be made by way of exposition rather than to be the object of solemn definition. And the Vatican Council simply repeats the words of Lateran, in which the expression *ab initio temporis* appears to be a mere explanatory incision.

However, the two Councils differ radically in scope.[5] The Vatican Council, in its teaching about God the Creator, opposed pantheism, which fosters two heresies about creation: the eternity of the world and the necessity of creation, or rather of emanation. To overthrow the basic error of pantheism in all its manifestations, the Council proclaims that creation was effected in time, and stresses God's complete freedom in creating. The purpose of eradicating pantheistic infiltrations directs the attention of the Fathers of the Council to these considerations and sheds new light on the formula, *ab initio temporis.*[6]

Consequently the doctrine that the universe has a temporal duration is defined as a dogma of faith. This judgment, common to all theologians, is confirmed by Pius XII in the encyclical *Humani generis* which, among many other errors listed and condemned, specifies the assertion that the world had no beginning, and adds that such a contention is contrary to the teaching of the Vatican Council.[7]

---

2. H. Denzinger, *Enchiridion symbolorum* (31st. ed., 1957) no. 428; *The Church Teaches* (St. Louis: B. Herder, 1955) no. 306.
3. Denzinger, no. 1783; *The Church Teaches,* no. 356.
4. Cf. A. D. Sertillanges, O.P., *L'Idée de création et ses retentissements en philosophie* (Paris: Aubier, 1945) p. 19.
5. Cf. P. De Haes, "Creatio in tempore," *Collectanea Mechliniensia,* XXXVI (1951) 587.
6. This is clear from the Acts of the Council: "The teaching about creation at the beginning of time is directly opposed to the eternal evolution or emanation asserted by pantheists." Cf. *Acta et decreta sacrorum conciliorum recentiorum,* in *Collectio lacensis,* VII, col. 520, adnot. 5
7. Encycl. *Humani generis* of Aug. 12, 1950, *Acta Apostolicae Sedis* 42 (1950) 570. The passage is contained in Denzinger (31st ed.) no. 2317, and in *The Church Teaches,* no. 364. For some other declarations of the magisterium of the Church touching this point, see Denzinger, 31, 501, ff., 706; *The Church Teaches,* 337 ff., 343.

II. Philosophy And the Problem of Eternal Creation

This dogma of faith is no obstacle to scientific exploration. Revelation teaches that the world began, but does not date that beginning; science has unimpeded liberty to search for the initial state from which the universe took its origin. Likewise, the dogma offers no embarrassment to philosophical reflection. The question of the possibility of creation from eternity and the question of the possibility of solving this problem by reason alone, remain legitimate objects of philosophical inquiry. Philosophers are free to speculate whether rational procedures are able to demonstrate the necessity of a temporal origin of the world, or whether they leave a choice between a universe that is finite in time and a universe existing and evolving in a limitless past.

Although the disputation on the eternity of the world challenges the attention of both philosophers and theologians, medieval Scholastics, like their predecessors in antiquity, usually treated the subject from a philosophical point of view. In the philosophical inquiry, whether the universe had a beginning in time or was created from eternity, the formula "in time" does not mean "in the midst of time," as though time had been flowing on prior to creation.[8] Since time is the measure of motion, there could have been no time before the existence of beings subject to motion, that is, creatures. Furthermore, the Scholastics understood the term "eternal" in an analogical sense. They did not debate whether the world could be eternal in the way that God is eternal. Eternity in the strict sense that is predicable of God alone is *tota simul,* simultaneously whole;[9] it excludes beginning, end, and all succession. In a wider sense that is pertinent to the present discussion, eternity excludes beginning or end or both, but not succession. Since the world has successive existence, the question is whether eternity is a communicable perfection somewhat in the way that God's goodness, wisdom, beauty and other attributes are communicable to creatures; concretely, whether the world could have existed always, without any beginning of its duration.

A correct idea about the beginning of the world is impossible unless it is preceded by a correct idea of creation from nothing. The problem of creation as such is separable from the question of an absolute beginning of things in time, and can be solved by human reason. Philosophy reaches a solution, not by proving a beginning of the world, but by considering the universe at the present hour or at any of its

---

8. St. Augustine, *De civitate Dei,* XI, 6 (PL 41, 322) had clarified this point: "The world was not made in time, but along with time."

9. The definition formulated by Boethius, *De consolatione philosophiae,* V, prosa 6 (PL 53, 858), "Eternity is the simultaneously whole and perfect possession of interminable life," was commonly accepted and approved by medieval philosophers and theologians.

moments in the past. It apprehends in the universe evidence of a basic dependence; the world cannot exist except in permanent reference to a Creator who causes it and sustains it in being. The world is not sufficient to itself; its changeableness is the first witness of its indigence. States succeed one another; none of them exists with absolute necessity: they can exist or not exist. The act of existing possessed by things depends at every instant on the influence of the cause of their existence. This changeableness, perceptible from the flights of galaxies down to the minute transformations which call forth emissions of energy in the cores of atoms, is a sure sign of contingence in the eyes of the metaphysician. The interdependence of beings is another indication of their ontological fragility; the elements of the universe act and react on one another. In such interdependence, increasingly recognized by science the farther it penetrates, the philosopher discerns a higher relationship, to the Being that depends on nothing.

This relationship of total dependence on God for existence is the very essence of creation. Whether that dependence has a beginning or lacks a beginning is quite accessory. Reason does indeed perceive that the world is not eternal with an uncreated eternity. But whether it can know by its own powers that the world was created in time is a completely different matter. This problem is closely connected with the philosophical controversy about the possibility of creation from eternity. If creation from eternity can be shown to involve a contradiction, creation had to have a beginning. But if the eternal duration of the world does not involve a contradiction, human reason is confronted with a formidable difficulty in its effort to demonstrate that it actually began. An a priori demonstration is ruled out, for God's creative will, from which such argumentation would have to be drawn, is free and therefore can be known by us only if God chooses to make it known by revelation. Attempts at a posteriori demonstration encounter the obstacle that the essence of created things, on which the proof would be based, abstracts from the notion of time. Conclusive arguments can be adduced to establish the fact that particular states of the world, such as its igneous state, or the expansion of the universe cannot be eternal, but do not show that its substance was not created from eternity, because the substance is independent of time, which is an accident.

Nevertheless, authors disagree on this point, now as well as formerly. Some assert that the beginning of the universe can be demonstrated by reason. Among them, St. Anselm, Richard of St. Victor, William of Paris, St. Bonaventure, and Henry of Ghent hold that a world without beginning is intrinsically repugnant, and hence that its finite duration is demonstrable. Of like mind are Alexander of Hales, Robert Grosseteste, Roger Bacon, John Peckham, Matthew of Aquasparta, Richard Middleton, Francis Toletus, and Gregory of Valencia. Opposed to this view is the contention that eternal creation or a universe existing without a first moment implies no intrinsic contradiction and hence is possible. This is the conviction of St. Thomas and his school, such as Giles of Rome, Capreolus, Cajetan, John of St. Thomas, and Dominic Banes, along with Louis de Molina, Francis Suarez, and a host of others.

III. Survey of the Philosophical Controversy on the Eternity of the World

Clarification of the relations between the idea of creation and that of beginning will be promoted by a glance at the origin of the conceptions respecting this problem. After primitive peoples had lost the primordial revelation concerning the one God and the creation of the universe, they turned their minds toward the fabrication of extravagant cosmogonies intermingled with fantastic theogonies. When the first philosophers began to investigate the cause of things, they not only repudiated such fables, but relegated the very origin of the world to the realm of fiction. Relying on the principle which seemed obvious to them, that nothing comes from nothing, *ex nihilo nihil fit,* they contended that the material substratum of the world is eternal.

Creation itself was generally regarded in antiquity as a process of arranging the elements of a primeval chaos whose causes were not sought, largely for the reason that a primal Necessity was envisioned as enveloping both God and the matter presupposed as requisite for His action. Eternal necessity was attributed not only to God, whose transcendence was unrecognized, but to a grand, divine All. God might be the supreme representative of this All, but He was not known simply as the first Being. The necessary All was the theater of phenomena that were open to examination. Thus the Ionians, philosophers and physicists, inquired into the causes of superficial changes occurring in bodies, and Plato, followed by Aristotle, studied the cause of substantial transformations that affected the innermost depths of things. But to all of them the problem presented itself on the basis of a common, pre-existing matter, which had no need of explanation outside itself. This matter is necessary; God is its witness, not its cause. God moves or activates the universal machine, but does not create it, for His causality presupposes something which is as necessary as He is, something which has the function of eternal passivity subect to the eternal influence of His activity.[10]

Plato's position on the formation of the world is quite clear. Since the universe is visible, tangible, and material, it was created, and consequently had to have a cause. Its father and maker fashioned it

10. Cf. A. D. Sertillanges, O.P., *L'Idée de création et ses retentissements en philosophie,* p. 6.
11. *Timaeus* (27 D - 29 B).

according to an eternal, unchangeable pattern, so that it is perfect.[11] Yet
God, the supremely intelligent artificer, did not create it out of nothing.
Because He is good, He desired that all things should resemble Him as
closely as possible. Therefore, finding that all visible matter was in a
chaotic condition of turbulent commotion, He reduced it to order from its
previous state of disorder.[12] Along with the heavens that were set in
orderly motion by the creator, time itself, which is the moving image of
unmoving eternity, first came into being.[13]

This view was unique in antiquity. Aristotle asserts that Plato
was the only one who held that time was created, and thoroughly dis-
agrees with him. For time is inconceivable apart from the moment, which
is the beginning of future time and the end of past time. Hence, there
must always be time before and after any moment. And if time is
limitless, motion, too, must be eternal, because time is the measure of
motion and is itself a kind of motion.[14] Moreover, motion necessarily
implies the existence of things that are movable, indeed, of things that are
moving. Therefore, things in actual motion have existed eternally.[15]

In fact, matter itself, as potentiality, is necessarily outside the
range of becoming. For if it ever began, something else must have existed
as the primary substratum from which it took its origin; but that is
impossible, because matter is precisely the primary substratum of
everything.[16] Furthermore, heavenly bodies, which move in circular
motion, were ungenerated, for circular motion has no contrary, and it is
contraries that cause generation as well as corruption.[17] Consequently
matter as potentiality and the heavenly bodies must be eternal, for
everything that is ungenerated is eternal.[18] The same is true of heaven
taken in its entirety. It never had a beginning, but is one and eternal, with
a duration that lacks start or finish, and embraces within itself the whole
of infinite time.[19] Accordingly, the universe as an orderly whole or
cosmos is eternal, and has been set in motion by a first unmoved Mover,
an eternal, substantial, actual Being.[20]

The teachings of these eminent masters, as well as Neoplatonist
emanation theories which are hardly reconcilable with the idea of creation
in time, had to be kept in mind by philosophers who had been imbued
with a doctrine of faith about the temporal beginning of the world

---

12. *Ibid.* (29 E-30 C).
13. *Ibid.* (37 C - 38 B).
14. Aristotle, *Phys.,* VIII, 1 (251b 10-28).
15. *Ibid.* (251a 9-b 9).
16. *Ibid.,* I, 9 (192a 25-34)
17. *De caelo,* I, 3 (270a 12-22).
18. *Ibid.,* I, 12 (283a 21-bI).
19. *Ibid.,* II, 1 (2823b 26-284a 2).
20. *Metaph.,* XII, 7 (1072a 21-26).

through divine creation. Such were Alexandrian and medieval Jews, many Fathers of the Church, Arabian philosophers, and Christian Scholastics.

Outstanding among the Alexandrian Jews is Philo, who generally follows the philosophical lead of Plato, although at times he borrows also from the Stoics. In his exegesis of the Old Testament he favored an excessively allegorical method, with the result that his tenets are not always clear. He is emphatic in teaching that the universe had a beginning and that it was brought into existence by God.[21] Whether he also taught the eternity of matter and hence was ignorant of creation from nothing, or held the creation of matter itself from no pre-existing substratum, is still debated among his interpreters.[22] The expressions he uses, such as the statement that God is not merely an artificer but the world's Creator, because He "made things which before were not,"[23] or that "nothing comes into being out of nonbeing,"[24] do not completely settle the controversy. In the light of recent study, however, the view that he maintained the very production of matter from nothing by God's creative causality appears to be highly probable, if not certain.[25] At any rate, there was no time before there was a world, for the conception of time depends on the movement of material bodies; hence time began either simultaneously with the cosmos or immediately thereafter.[26] Time is, therefore, inseparable from the motion of the universe and was created by God, as the cosmos itself was.

With the advent of Christianity the idea of creation as production of being from nothing became a commonplace. Total dependence of all finite reality on the first Cause was perceived to be its essential element: the notion of duration came to be recognized as secondary. The question was this: Does God's creative action suppose something pre-existing or does it suppose nothing? Either something escapes God's causality or else God is the Author of all contingent being, including the material substratum of being. According to the response, the domain of God is extended or restricted, so as to manifest truly or to compromise irremediably the knowledge, power, and authority of the first Principle. The Fathers of the Church had an intense appreciation of the importance of the issue. Relying on Scriptural evidence, they took so firm a stand that some of them ran the risk of confusing the idea of creation in its essentials with the idea of beginning in time, which is only one of its modalities, or rather is a question of fact which, though unmistakably

---

21. Philo, *De opificio mundi*, 2, 7-12; *De fuga et inventione,*2, 12.
22. Cf. H. A. Wolfson, *Philo: Foundations of Religious Philosophy in Judaism, Christianity, and Islam* (Cambridge, Mass.: Harvard University Press, 1947) I, 300f.
23. Philo, *De somniis*, I, 13, 76.
24. *De aeternitate mundi*, 2, 5.
25. Wolfson, *op. cit.*, pp. 303-10, 323.
26. Philo, *De opificio mundi*, 7, 26-28; *De somniis*, I, 32, 187.

showing forth the dependence of the world on God and the complete freedom of God in creating, is not indispensable for religious life.

Patristic testimony on the finite duration of the universe is unanimous. Occasionally, in the heat of controversy against pagan philosophers or Gnostics who asserted that matter was uncreated and coeternal with God, some Fathers went beyond the question of fact, which alone is revealed, and contended that eternal creation was impossible. Particularly in their polemic against the Arians, several of them argued that the Son of God was not a creature, as the heretics held, because He was eternal and therefore could not be created but must be divine. According to Tertullian, eternity belongs to God alone, because it is His special property; if it were attributed to some other being as well, it would no longer be a special prerogative of God but would be shared with that other being. "God must be One, because what is supreme is God; but nothing is supreme except what is unique; and nothing can be unique if something else is put on a par with it; but matter is put on a par with God if it is held to be eternal."[27] St. Athanasius insists: "Things which have been made cannot be eternal, even though God could always have made them. For they came from nothing, and did not exist before they were made. But how could things which did not exist before they were made, be co-existent with God who always is?"[28] Writing against Neoplatonists in the sixth century, Zacharius Rhetor, Monophysite Bishop of Mitylene, echoes the same views: "The world is not co-eternal with God; for everything that has been made is subsequent to its maker both in causality and in time."[29] Even St. Augustine seems to doubt that creation from eternity is possible; the Neoplatonist position that the universe, though created, has an eternal existence, is to him "scarcely intelligible."[30]

The problem of the eternity of the world was often debated among Arabian theologians and philosophers. The latter followed Aristotle so zealously that they were unwilling to depart from any doctrine which they thought he held; however, since they were acquainted with the teaching on creation proposed in the Koran, they were better equipped to explain the influence of the first Mover on matter, which is obscure in Aristotle. Moreover, under the influence of the Neoplatonist theory of emanation, they taught that God was not only the first Mover of the universe, but also the Creator from whom the world eternally emanates.

---

27. *Adversus Hermogenem*, 4 (PL 2, 201).
28. *Adversus Arianos orationes*, 1, 29 (PG 26, 72).
29. *De mundi opificio* (PG 85, 1113).
30. *De civitate Dei*, XI, 4 (PL 41, 319). Cf. J. de Blic, "Les arguments de saint Augustin contre l'éternité du monde," *Mélanges de Science Religieuse*, II (1945), 33-34.

Avicenna, one of the great names in the history of philosophy, distinguishes between eternity according to time and eternity according to essence. The former pertains to the world, which had no beginning in time but has existed during infinite past time. The latter pertains to God, whose essence is uncaused.[31] Prime matter and incorruptible forms are eternal.[32] Motion likewise is eternal; it has a beginning in the sense that it proceeds from the Creator, but had no beginning in time.[33] Hence the heaven is eternal, for its circular motion has no contrary that could cause any change in its disposition; therefore it is incapable either of beginning or ending.[34]

Conflict was inevitable between philosophers who entertained such views and orthodox Mohammedan theologians, the Mutakallimun ("dialecticians") who, relying on the Koran,[35] defended the "newness" of the world, that is, its origin from nothing by God's creation at the beginning of time. Algazel, Islam's most prominent theologian, attacked the philosophers in his defense of the faith, *Destructio philosophorum*, the longest portion of which deals with the eternity of the world, the problem he considers to be the most important. Against him, Averroes, last of the great philosophers of Islam and the most scholarly commentator of Aristotle, wrote his *Tahafut al-Tahafut*, that is, *The Incoherence of the Incoherence.*[36] In the entire "First Discussion," Averroes upholds the eternity of the world and of time, with a detailed reply to the assaults of his adversary.[37]

About the same time as the Arabians, Jews in Babylon and in Spain began to discuss philosophical problems, and undertook to defend their main religious doctrines against both philosophers and liberal theologians. The most eminent of them in genius and influence was Moses Maimonides, whose *Moreh Nebuchim* or *Guide for the Perplexed* is a sort of summa of Jewish theology, with philosophical inspiration derived from Neoplatonism and particularly from Aristotelianism. Confident that harmony between philosophy and his beloved Prophets was possible, he addressed his work to those who were acquainted with philosophy and the

---

31. *De diffinitionibus et quaesitis* (text in M. Gierens, S.J., *Controversia de aeternitate mundi*, in *Textus et Documenta*, Series philosophica, 6 [Romae: Pontificia Universitas Gegoriana, 1933] no. 30).

32. *Sufficientia*, I, 3 (Gierens, no. 27); *Metaphysics compendium* , tr. 6, 2 (ed. N. Carame, p. 56).

33. *Metaphys.*, tr. 9, 1 (Gierens, no. 29).

34. *De caelo et mundo*, c.4 (Gierens, no. 28).

35. Suras 7, 52; 10, 3; 11, 9; 32, 3, in A. J. Arberry, *The Koran Interpreted* (London, New York: Macmillan, 1955).

36. The Latin version of the title, *Destructio destructionis* (or *destructionum*) is inexact; cf. Averroes' *Tahafut al-Tahafut*, translated from the Arabic with Introduction and Notes by S. Van den Bergh (London: Luzac, 1954) I, xiii.

37. *Ibid.*, pp. 1-69.

sciences, but were at a loss how to reconcile their knowledge with the teachings of Scripture.

On the basis of the Old Testament, Maimonides maintains firmly that the universe was brought by God into existence out of non-existence. In the beginning was God alone, and nothing else. God produced from nothing all that exists, including time itself; for time depends on motion, and the things upon whose motion time depends are themselves created beings that have passed from non-existence into existence.[38] If the problem is considered from the philosophical viewpoint, it is insoluble. The dialectical arguments brought forward by the Mutakallimun to prove creation are inconclusive.[39] On the other hand, reason cannot prove the eternity of the world;[40] Aristotle himself was fully aware that his theory of the universe could not be demonstrated and that his arguments were no more than probable.[41]

Among the medieval Scholastics who submitted to the influence of Aristotle, none exceeded Siger of Brabant in enthusiasm. In his judgment, the Stagirite's teaching is the very voice of reason and philosophy. Although, as a Christian, he professed belief in the creation of the universe by God, as a philosopher he contended that Aristotle had demonstrated the eternity of the world so convincingly that his arguments could not be refuted by reason. Siger relies heavily on the consideration that the first Mover is always in act, not in potentiality prior to being in act; hence He always acts and always makes whatever He produces; consequently no species of being is actualized from previous potentiality, but was already in existence.[42] In the interest of self-protection, Siger prudently has recourse to the familiar device that he is merely reporting an opinion, without necessarily giving it his personal approval.[43]

Most of the eminent Scholastics of the thirteenth century, who accepted from revelation the fact that the world was created in time, maintained further that it could not have an eternal duration. Alexander of Hale employs an argument that turns up again and again: the very notion of creature involves possession of existence following upon non-existence, so that it must have a beginning of its duration.[44] Since an infinite regression in causality is impossible, in the domain of material causes we must come to one that is first. This first material cause does

---

38. *The Guide for the Perplexed,* translated from the original Arabic text by M. Friedländer (2nd ed.; reprinted London: G. Routledge, 1947) II, 13 (p. 171).
39. *Ibid.,* II, 16 (p. 178).
40. *Ibid.,* II, 16 and 17 (p. 178 ff.).
41. *Ibid.,* II, 15 (p. 176).
42. R. Barsotti, *Sigeri de Brabantia De aeternitate mundi,* in *Opuscula et Textus,* Series scholastica, 13 (Monasterii: Aschendorff, 1933) p. 26.
43. "Haec autem dicimus opinionem Philosophi recitando, non ea asserendo tamquam vera" (*ibid.*).
44. *Summa theologica,* I, inq. 1, tr. 2, q. 4, men. 2, cap. 4 (Quaracchi, I, no. 64, p. 95 b).

not produce itself; hence it comes into existence subsequent to complete privation, Accordingly, it is not eternal, and therefore the world is not eternal.[45] St. Albert, in his commentary on the *Sentences* , thinks that the non-eternity of the universe is more probable from the standpoint of philosophy.[46] Later he took a more positive stand and held that no creature, whether spiritual or corporeal, could have existed from eternity. The position of those who admit the creation of the world by God, yet teach that it is co-eternal with God in the sense that it has no beginning of its duration, is regarded as unintelligible.[47]

Of all the Scholastics who contend that creation in time is rationally demonstrable, the most emphatic is St. Bonaventure. He not only refutes the main arguments proposed by Aristotle, but takes the offensive to prove that the hypothesis of an eternal duration of a universe created by God is intrinsically repugnant. The contradiction is so patent to him that he does not believe that any philosopher, no matter how slight his intellectual competence, has ever held such a view.[48] An infinite series cannot be crossed. But if the world does not begin, an infinite number of days has preceded the present one. Therefore it is impossible to traverse them, and so we could never have arrived at today.[49] Moreover, if the world is eternal, the human race must also be eternal, because the universe, existing as it does for the sake of man, was never without men. Hence men in infinite number have existed, and their immortal, rational souls must be now existing. But a simultaneously infinite number is impossible. Consequently the world must have begun in time.[50] The very concept of creation implies a temporal beginning. What is created by God from nothing preexisting, is not derived from nothing regarded as a material substratum. Hence the formula *ex nihilo* indicates temporal succession. In created beings, therefore, non-existence precedes existence.[51]

IV. Position of St. Thomas

For St. Thomas Aquinas, the question of the eternity of the world was a highly current and vital problem. Although Aristotle was at that time dominating the philosophical thought of Europe, his thesis of the eternal universe was encountering violent opposition within the

---

45. *Ibid.,* II, inq. 1, tr. 2, tit. 4, cap. 1 (Quaracchi, II, no. 67, p. 85 b).
46. *In II Sent.,* d. 1, a. 10 (ed. Borgnet, XXVII, p. 28 a).
47. *Summa theol.,* II, tr. 1, q. 4, men. 2, a.5 (Gierens, no. 45).
48. *In II Sent.,* d. 1, par. 1, a. 1, q. 2 (Quaracchi, II, p. 22b).
49. *Ibid.,* fund. 3 (Quaracchi, II, p. 21a)
50. *Ibid.,* fund. 5 (Quaracchi, II, p. 21 b).
51. *Ibid.,* fund. 6 (Quaracchi, II, pp. 22 a).

Church, for it flatly contradicted the dogma of faith about creation in time. In the philosophical domain, too, the controversy was heated: up to the middle of the thirteenth century practically all the great Christian doctors firmly maintained that reason could demonstrate the necessity of a beginning for the world. The contrary view of the most famous among them soon reversed the philosophical trend.

From the earliest years of his literary productivity to the end of his life, St. Thomas turned again and again to the subject of the eternity of the world. He regarded the question as so important that he wrote a treatise on it, *De aeternitate mundi*. On six other occasions, in different works and at various periods of his career, he expressly took up the same theme, notably in the *Scriptum super IV libros Sententiarium magistri Petri Lombardi*, II, dist. 1, q. 1, a.5 (ca. 1256); *Summa contra Gentiles*, II, cc. 31-38 (ca. 1262); *De Potentia Dei* (1259/68), q.3, aa. 14, 17; *Summa theologiae* Ia, q. 46 (ca. 1265); *Quodlibetum* III, q. 14 (ca. 1270); and the *Compendium theologiae*, cc. 98 f. (ca. 1271).[52] His appreciation of the gravity of the issue is clear from his amplitude of treatment and careful method. Even in the *Summa theologiae* he treats the question thoroughly, as is evident from the many objections, much more numerous than usual, which he lists and undertakes to refute.

In most of his contributions to the controversy, St. Thomas proposes two questions. First, is it possible to prove the eternity of the world, and do the arguments of the "philosophers"[53] really demonstrate their thesis? Secondly, is it possible, as theologians of the age claimed, to prove with certainty the temporal origin of the universe?

With regard to the first question, St. Thomas recognized a certain suasive force in the arguments for the eternity of the universe as drawn up by Aristotle and later by the Arabian philosophers, yet denied that they were true demonstrations. On the second question, which is of greater interest, he took a resolute stand and maintained it consistently; reason can, indeed, demonstrate the world was created, but not that it was created in time. The view of imagination, that the world was made subsequent to nothingness, must be corrected by the intellect. Creation is not a passage from empty time to a time filled with successively existing beings, for there is no duration apart from beings that endure. Creation is not

---

52. M. D. Chenu, O.P. *Introduction à l'étude de saint Thomas d'Aquin* (Montreal, Paris Institut d'études médiévales, 1950) p. 283, thinks that the *Compendium* was written prior to the *Summa*, perhaps around 1265-67. See also I. T. Eschmann, O.P., *A Catalogue of St. Thomas' Works*, in E. Gilson, *The Christian Philosophy of St. Thomas Aquinas* (New York: Random House, 1956) p. 412.

53. According to M. D. Chenu, "Les philosophes' dans la philosophie chrétienne médiévale," *Revue des Sciences Philosophiques et Théologiques*, XXVI (1937) 28 ff., these philosophers are pagans or infidels who lack the supernatural light of faith and consequently have sought to understand the world and to organize human moral life with the sole resources of their reason.

change, but is a unilateral relation of dependence in the creature with respect to the Creator as cause of the creature's existence. Hence the idea of creation does not logically require beginning. The creature exists as an effect of God's creative act; if the Creator wills the creature to exist without inception of its duration, it so exists; if the Creator wills it to have a finite, limited duration, it has a beginning of its existence. Therefore creation is essentially dependence in being; if the notion of newness or beginning is associated with it, this is because of the fact acknowledged by Christians, not because of the very essence of creation. A universe without initial instant would still be a created universe dependent on the first Cause, and its successive unfolding in time would leave it infinitely inferior to God's eternity. Accordingly we need entertain no fear that a creature lacking beginning would rival God in duration, for no comparison is possible between time, even though unlimited, and the immutable possession of eternity. God alone is without beginning, end, and succession; He alone is changeless, intemporal.

The ultimate reason why a philosophical demonstration must turn out to be inconclusive is that the fact of temporal creation depends on God's free will, which cannot be known except by revelation. Since such a revelation has been communicated to us, we assert with complete assurance that the world has had a beginning; but our certain knowledge is based on faith alone.

Although a strict rational demonstration for the temporal inception of the world is not forthcoming, the mind can discern reasons of fittingness which are capable of illuminating our faith. Thus God's creative power is more strikingly apprehended by us if the world has not existed eternally but began in time. Again, the divine goodness shines forth more brilliantly to our dull eyes if the dependence and contingence of the world are manifested by its temporal origin. Other arguments may incline the mind to assent, but none of them can compel it; they have a relative value and are open to refutation.

In the solution invariably proposed by St. Thomas, therefore, no metaphysical necessity is discernible either for a beginning or for an indefinite past of the universe. Neither of the two alternatives involves contradiction; both leave the world in dependence on God. Whether the world began or not, it bears within itself the signs of its basic indigence, and requires the Creator's continuous creative activity or conservation of its being.

The arguments adduced by the so-called "Augustinian" theologians against the position defended by St. Thomas did not cause him any serious difficulty. He was able to point out that they are mostly sophisms. In his judgment they are weak, empty, or devoid of force; some of them are even ridiculous, and are of a nature to arouse scornful laughter in infidels. Only one objection occasioned him some embarrassment.

If the human race had no beginning, an infinite number of immortal, human souls must be in actual existence; the infinity is simultaneous, not successive. St. Thomas remarks that this question raises many others and that one could escape the difficulty in many ways. He lists the solutions that have been proposed, and notes that the objection is not pertinent, for it is limited to a particular case whose force comes from the special nature of the soul, not from the general notion of creature or infinite being. Even though the human race could not be eternal, the possibility of an eternal existence for the material world and the angels is not thereby excluded. His own purpose, abstracting from this particular question, is to investigate the general issue, whether some creature could exist eternally. In the opusculum, *De aeternitate mundi,* he adds, somewhat defiantly, that no one has yet demonstrated that God is unable to produce an actually infinite multitude.

However, this problem, which St. Thomas regarded as very delicate, has lost all practical importance for us, and indeed has completely vanished. We now know, quite apart from revelation, that man has not always been on earth, and that the earth itself is considerably younger than our galaxy, the Milky Way.

We may wonder why St. Thomas, who believed firmly, with all the certitude of faith, in the revealed truth that the world had a beginning of its time, persisted in seven different works on showing that such a beginning could not be rationally demonstrated. He certainly did not desire to create difficulties for apologetics by undermining serviceable arguments. But he was convinced that bad arguments could compromise the best of causes. In the environment of peripatetic philosophers who taught the eternity of the world, he feared that Catholics would expose themselves to ridicule if they provided infidels with an occasion for thinking that the truths of faith were accepted on insecure grounds. He resolutely opposed anyone who risked humiliating the faith by proposing unsound demonstrations of truths that are in reality incapable of demonstration.

In the thirteenth century scientific cosmogony was, of course, totally unknown; St. Thomas, like Aristotle before him, was necessarily ignorant of the scientific arguments worked out in recent years to prove the beginning of the universe. Until some demonstration eliminates the possibility of matter having existed, prior to the initial phase of the world's evolution, in a state beyond our power of investigating, or of a universe oscillating forever in alternating expansion and contraction,[54] the purely philosophical problem has an object even in the material universe. In any

---

54. Cf. C. Vollert, S.J., "Origin and Age of the Universe Appraised by Science," *Theological Studies,* XVIII (1957) 165 ff.

case, it persists with reference to the world of spiritual beings that were not unknown to Aristotle[55] and were well known to St. Thomas, the creatures we call angels.

The translations here presented disclose the mind of St. Thomas in the celebrated debate on the philosophical possibility of an eternal universe, and show how steadfastly he stood his ground from his earlier years to the close of his life. Of these five selections, the least known is perhaps the brief treatise *On the Eternity of the World*. The words, *contra murmurantes,* sometimes added to the title, *De aeternitate mundi,* are not found in the older manuscripts.[56] Recent research has shown that the opusculum was not written in 1270, the year usually assigned, but belongs to the period preceding Part I of the *Summa theologiae;* the exact year cannot be determined.[57] The work seems to be a public lecture directed against certain conservative theologians of the "Augustinian" school. In his *Scriptum super libros Sententiarum* as well as in the *Contra Gentiles,* St. Thomas had argued that the beginning of the world could not be demonstrated but was a truth knowable only through revelation. This contention could hardly have failed to arouse opposition in the camps of overorthodox theologians who were convinced that reason could prove the temporal origin of the universe. St. Thomas apparently perceives in the controversy an opportunity to present his views on a matter of principle, the distinction to be made between questions of faith and those which offer room for freedom of investigation. Because of the seriousness of the issue, he seems at times to depart from his usual serenity and to experience some irritation at the lack of perception and the attacks of his opponents. This may account for his precise statement of the question and his polemical severity. In addition to its intrinsic merits, the opusculum is important for the insight it gives us into Aquinas the controversialist.[58]

In translating the *De aeternitate mundi* I used the text established by J. Perrier, O.P., *St. Thomae Aquinatis Opuscula omnia necnon opera minora* (Paris: P. Lethielleux, 1949) I, 53-61, and also consulted the one edited by R. M. Spiazzi, O.P., *Divi Thomae Aquinatis Opuscula philosophica* (Taurini, Romae: Marietti, 1954) pp. 105-108. The two chapters from the *Compendium theologie* are here reprinted, with the kind permission of the publishers, from my translation of that work, *Compendium of Theology* (St. Louis: B. Herder Book Co., 1948) pp. 98-103.

---

55. *Metaph. XII, 8* (1073a 14-38).
56. F. Pelster, S.J., "Zur Echtheit der Concordantia dictorum Thomae und zur Datierung von De aeternitate mundi," *Gregorianum,* XXXVII (1956), 619.
57. *Ibid.,* p. 622.
58. Cf. M. D. Chenu, *Introduction à l'étude de saint Thomas d'Aquin,* p. 289f.

St. Thomas Aquinas

On the Eternity of the World
(De Aeternitate Mundi)

1. If we suppose, in accord with Catholic faith, that the world has not existed from eternity but had a beginning of its duration, the question arises whether it could have existed forever. In seeking the true solution of this problem, we should start by distinguishing points of agreement with our opponents from points of disagreement.

If the question is phrased in such a way as to inquire whether something besides God could have existed forever, that is, whether a thing could exist even though it was not made by God, we are confronted with an abominable error against faith. More than that, the error is repudiated by philosophers, who avow and demonstrate that nothing at all can exist unless it was caused by Him who supremely and in a uniquely true sense has existence.

However, if we inquire whether something has always existed, understanding that it was caused by God with regard to all the reality found in it, we have to examine whether such a position can be maintained. If we should decide that this is impossible, the reason will be either that God could not make a thing that has always existed, or that the thing could not thus be made, even though God were able to make it. As to the first alternative, all parties are agreed that God could make something that has always existed, because of the fact that His power is infinite.

2. Accordingly our task is to examine whether something that is made could have existed forever. If we reply that this is impossible, our answer is unintelligible except in two senses or because there are two reasons for its truth: either because of the absence of passive potentiality, or because of incompatibility in the concepts involved.

The first sense may be explained as follows. Before an angel has been made, an angel cannot be made, because no passive potentiality is at hand prior to the angel's existence, since the angel is not made out of pre-existing matter. Yet God could have made the angel, and could also have caused the angel to be made, because in fact He has made angels and they have been made. Understanding the question in this way, we must simply concede, in accordance with faith, that a thing caused by God cannot have existed forever, because such a position would imply that a passive potentiality has always existed, which is heretical. However, this does not require the conclusion that God cannot bring it about that some being should exist forever.

Taken in the second sense, the argument runs that a thing cannot be so made because the concepts are incompatible, in the same way as affirmation and denial cannot be simultaneously true; yet certain people assert that even this is within God's power. Others contend that not even God could make such a thing, because it is nothing. However, it is clear that He cannot bring this about, because the power by which it is supposed to be effected would be self-destructive. Nevertheless, if it is alleged that God is able to do such things, the position is not heretical, although I think it is false, just as the proposition that a past event did not take place involves a contradiction. Hence Augustine, in his book against Faustus, writes as follows: "Whoever says, 'If God is omnipotent, let Him bring it about that what has been made was not made,' does not perceive that what he really says is this: 'If God is omnipotent, let Him bring it about that what is true is false for the very reason that it is true.'"[1] Still, some great masters have piously asserted that God can cause a past event not to have taken place in the past; and this was not esteemed heretical.

3. We must investigate, therefore, whether these two concepts are logically incompatible, namely, that a thing has been created by God and yet has existed forever. Whatever may be the truth of the matter, no heresy is involved in the contention that God is able to bring it about that something created by Him should always have existed. Nevertheless I believe that, if the concepts were to be found incompatible, this position would be false. However, if there is no contradiction in the concepts, not only is it not false, but it is even possible; to maintain anything else would be erroneous. Since God's omnipotence surpasses all understanding and power, anyone who asserts that something which is intelligible among creatures cannot be made by God, openly disparages God's omnipotence. Nor can anyone appeal to the case of sin; sins, as such, are nothing.

The whole question comes to this, whether the ideas, to be created by God according to a thing's entire substance, and yet to lack a beginning of duration, are mutually repugnant or not. That no contradiction is involved, is shown as follows. A contradiction could arise only because of one of the two ideas or because of both of them together; and in the latter alternative, either because an efficient cause must precede its effect in duration, or because non-existence must precede existence in duration; in fact, this is the reason for saying that what is created by God is made from nothing.

4. Consequently, we must first show that the efficient cause, namely God, need not precede His effect in duration, if that is what He Himself should wish.

---

1. *Contra Faustum*, XXVI, 5 (PL, 42, 481).

In the first place, no cause producing its effect instantaneously need precede its effect in duration. Now God is a cause producing His effect, not by way of motion, but instantaneously. Therefore He need not precede His effect in duration. The major premise is clear from induction, based on all instantaneous changes, such as illumination, and the like. It can also be demonstrated by reasoning. In any instant in which a thing is asserted to exist, the beginning of its action can likewise be asserted, as is evident in all things capable of generation; the very instant in which fire begins to exist, it emits heat. But in instantaneous action, the beginning and the end of the action are simultaneous, or rather are identical, as in all indivisible things. Therefore, at any moment in which there is an agent producing its effect instantaneously, the terminus of its action can be realized. But the terminus of the action is simultaneous with the effect produced. Consequently no intellectual absurdity is implied if we suppose that a cause which produces its effect instantaneously does not precede its effect in duration. There would be such an absurdity in the case of causes that produce their effects by way of motion, because the beginning of motion must precede its end. Since people are accustomed to think of productions that are brought about by way of motion, they do not readily understand that an efficient cause does not have to precede its effect in duration. And that is why many, with their limited experience, attend to only a few aspects, and so are overhasty in airing their views.

This reasoning is not set aside by the observation that God is a cause acting through His will, because the will, too, does not have to precede its effect in duration. The same is true of the person who acts through his will, unless he acts after deliberation. Heaven forbid that we should attribute such a procedure to God!

5. Moreover, the cause which produces the entire substance of a thing is no less able to produce that entire substance than a cause producing a form is in the production of the form; in fact, it is much more powerful, because it does not produce its effect by educing it from the potentiality of matter, as is the case with the agent that produces a form. But some agent that produces only a form can bring it about that the form produced by it exists at the moment the agent itself exists, as is exemplified by the shining sun. With far greater reason, God, who produces the entire substance of a thing, can cause His own effect to exist whenever He Himself exists.

Besides, if at any instant there is a cause with which the effect proceeding from it cannot co-exist at that same instant, the only reason is that some element required for complete causality is missing; for a complete cause and the effect caused exist together. But nothing complete has even been wanting in God. Therefore an effect caused by Him can exist always, as long as He exists, and so He need not precede it in duration.

Furthermore, the will of a person who exercises his will suffers no loss in power. But all those who undertake to answer the arguments by which Aristotle proves that things have always had existence from God for the reason that the same cause always produces the same effect,[2] say that this consequence would follow if He were not an agent acting by His will. Therefore, although God is acknowledged to be an agent acting by His will, it nevertheless follows that He can bring it about that what is caused by Him should never have been without existence.

And so it is clear that no logical contradiction is involved in the assertion that an agent does not precede its effect in duration. As regards anything that does imply logical contradiction, however, God cannot bring it into being.

6. We now proceed to inquire whether logical contradiction is latent in the position that a created thing was never without existence. The reason for doubting is that, since such a thing is said to have been made from nothing, non-existence must seemingly precede its existence in the order of duration. The absence of any contradiction is shown by Anselm in the eighth chapter of his *Monologium*, where he explains how a creature may be said to have been made from nothing. "The third interpretation," he states, "according to which something is said to have been made from nothing, is reasonable if we understand that the thing was, indeed, made, but that there is nothing from which it was made. In a like sense we may say that, when a man is saddened without cause, his sadness arises from nothing. In this sense, therefore, no absurdity will follow if the conclusion drawn above is kept in mind, namely, that with the exception of the supreme essence all things that exist were made by it out of nothing, that is, not out of something."[3] According to this explanation, then, it is clear that no order is established between what was made and nothing, as though what is made would first have to be nothing, and would afterward be something.

7. To proceed further, let us suppose that the order alluded to above, namely, relationship to nothingness, remains asserted, so that the sense is this: the creature is made from nothing [*ex nihilo*] that is, it is made after nothing. The term "after" unquestionably connotes order. But order is of various kinds; there is an order of duration and an order of nature. If, therefore, the proper and the particular do not follow from the common and the universal, it will not be necessary, just because the creature is said to exist subsequent to nothingness, that it should first have been nothing, in the order of duration, and should later be something. It

---

2. *Phys.*, III, 4 (203b 27-30).
3. *Monologium*, 8 (PL, 158, 156; ed. Schmitt, I, 23).

is enough that in the order of nature it is nothing before it is a being; for
that which befits a thing in itself is naturally found in it before that which
it merely has from another. But a creature does not have existence except
from another; regarded as left simply to itself, it is nothing; prior to its
existence, therefore, nothingness is its natural lot. Nor, just because
nothingness does not precede being in duration, does a thing have to be
nothing and being at the same time. For our position is not that, if the
creature has always existed, it was nothing at some time. We maintain
that its nature is such that it would be nothing if it were left to itself; just
as, if we say that the air was always illuminated by the sun, we must hold
that the air has been made luminous by the sun. And because everything
that comes into being comes from what is not contingent, that is, from
that which does not happen to exist along with that which is said to
become, we must assert that the air was made luminous from being not
luminous or from being dark; not in the sense that it was ever not-
luminous or dark, but in the sense that it would be such if it were left to
itself alone. And this is brought out more clearly in the case of stars and
planets that are always being illuminated by the sun.

8. Thus it is evident that the statement that something was made
by God and nevertheless was never without existence, does not involve
any logical contradiction. If there were some contradiction, it is surprising that Augustine did not perceive it, as this would have been a most
effective way of disproving the eternity of the world; and indeed he brings
forward many arguments against the eternity of the world in the eleventh
and twelfth books of *De civitate Dei* ;[4] yet he completely ignores this line
of argumentation. In fact, he seems to suggest that no logical contradiction is discernible here. Thus in Book X, chap. 31 of *De civitate Dei,* he
says of the Platonists: "They found a way of accounting for this by
explaining that it was not a beginning of time but a principle of subordina-
tion. They point out that if a foot had always, from eternity, been planted
in the dust, there would always be a footprint underneath, and no one
would doubt that the footprint had been made by someone stepping
there; and yet the foot would not be prior to the print, although the print
was made by the foot. In the same way, they continue, the world and the
gods created in it have always existed, since He who made them has
always existed, and nevertheless they were made."[5] Augustine never
charges that this is unintelligible, but proceeds against his adversaries in
another way. He also says, in Book XI, chap. 4: "They who admit that the
world was made by God, yet do not wish it to have a beginning in time but
only a beginning of its creation, so that it was always made in some sense

---

4. *De civ. Dei,* XI, cc. 4-6; XII, cc. 15, 16 (PL, 41, 319 ff.; 364 f.).
5. *Ibid.,* X, 31 (PL, 41, 311).

that is scarcely intelligible, do indeed say something."[6] How and why this
is scarcely intelligible, was touched on in the first argument.[7]

9. Another surprising thing is that the best philosophers of
nature failed to discern this contradiction. In the fifth chapter of the same
book, Augustine, writing against those who were mentioned in the
preceding reference, remarks: "Our present discussion is with those who
agree with us that God is incorporeal and is the Creator of all natures,
with the exception of His own." And, regarding the latter, he adds,
further on: "They surpassed all other philosophers in prestige and
authority."[8] The same situation emerges if we carefully consider the position of those who held that the world has always existed; for in spite of
this they teach that it was made by God, and perceived no logical inconsistency in this doctrine. Therefore they who do descry such inconsistency
with their hawk-like vision are the only rational beings, and wisdom was
born with them!

10. Yet, since certain authorities seem to be on their side, we
have to show that the foundation furnished by these authorities is fragile.
Damascene, for instance, in the eighth chapter of the first book, observes:
"What is brought to existence from non-existence is not of such a nature
as to be co-eternal with Him who is without beginning and exists
forever."[9] Similarly Hugh of St. Victor says, in the beginning of his book,
De sacramentis : "The ineffable power of omnipotence could not have
anything co-eternal with it, so as to have aid in creating."[10]

The minds of these authorities and of others like them is clarified
by what Boethius says in the last book of the Consolation: "Certain people, when they learn about Plato's view that this world did not have a
beginning in time and is to have no end, wrongly conclude that the created world is thus made co-eternal with its Creator. But it is one thing to
be carried through an endless life, which is what Plato attributed to the
world, and quite another to embrace the whole presence of endless life all
at once, which is manifestly proper to the divine mind."[11]

11. Hence it is clear that the difficulty feared by some does not
follow, that is, that the creature would be on a par with God in duration.
Rather we must say that nothing can be co-eternal with God, because
nothing can be immutable save God alone. The statement of Augustine
in De civitate Dei, XII, chap. 15, is to the point: "Since the flight of time
involves change, it cannot be co-eternal with changeless eternity.
Accordingly, even though the immortality of the angels does not run on in

6. *Ibid.,* XI, 4 (PL, 41, 319).
7. Cf. no. 4 above.
8. *De civ. Dei,* XI, 5 (PL, 41, 320 f.).
9. St. John Damascene, *De fide orthodoxa,* I, 8 (PG, 94, 814).
10. Hugh of St. Victor, *De sacramentis,* I, 1 (PL, 176, 187).
11. Boethius, *De consolatione philosophiae,* V, pr. 6 (PL, 63, 859).

time, and is not past as though it were no longer present, or future as though it had not yet arrived, yet their movements, by which successive times are traversed, do change over from the future into the past. And therefore they cannot be co-eternal with the Creator, in whom we cannot say that any movement has occurred that no longer endures, or that any will occur that has not yet taken place."[12] He speaks in like vein in the eighth book of his commentary on Genesis: "Because the nature of the Trinity is absolutely changeless, it is eternal in such a way that nothing can be co-eternal with it."[13] And he utters similar words in the eleventh book of the *Confessions*.[14]

12. They also bring in arguments which philosophers have touched on, and then undertake to solve them. One among them is fairly difficult; it concerns the infinite number of souls: if the world has existed forever, the number of souls must now be infinite. But this argument is not to the purpose, because God could have made the world without men and souls; or He could have made men at the time He did make them, even though He had made all the rest of the world from eternity. Thus the souls surviving their bodies would not be infinite. Besides, no demonstration has as yet been forthcoming that God cannot produce a multitude that is actually infinite.

There are other arguments which I forbear to answer at the present time. A reply has been made to them in other works.[15] Besides, some of them are so feeble that their very frailty seems to lend probability to the opposite side.

---

12. *De civ. Dei,* XII, 15 (PL, 41, 364).
13. *De Genesi ad litteram,* VIII, 23 (PL, 34, 389).
14. *Confessiones,* XI, 30 (PL, 32, 826).
15. See responses to objections in the selections from *Contra Gentiles, De potentia,* and *Summa theologiae* included in this book.

St. Thomas Aquinas

Summa Contra Gentiles

Book II, Chapters 31-38

Chapter 31

That Creatures Need Not Have Existed Always

We have still to show that created things need not have existed
from eternity

If the entire created universe or any single creature necessarily
exists, this necessity must arise either from the creature itself or from
some other being. The necessity cannot be derived from the creature
itself. For we showed above that every being must proceed from the first
Being.[1] But anything that does not enjoy self-existence cannot possibly
derive necessary existence from itself, because what exists necessarily cannot lack existence; and thus that which of itself has necessary existence,
also has of itself the impossibility of not being. Consequently it cannot be
a non-being; hence it is a being.

If, however, the necessity in question is derived from some other
being, it must come from some cause that is extrinsic, because everything
that is received within a creature owes its existence to another. Now an
extrinsic cause is either efficient or final. The effect of an efficient cause
exists necessarily when the agent acts necessarily, for the effect depends
on the efficient cause in consequence of the agent's action. Hence, if the
agent is not constrained by any necessity to the action of producing the
effect, the effect does not have to follow with absolute necessity. But God
is not compelled by any necessity to the action of producing creatures, as
was shown above.[2] Therefore, as far as the necessity that depends on an
efficient cause is concerned, no creature has to exist with absolute necessity.

The same is true of the necessity that depends on a final cause.
For things that are means to an end do not acquire necessity from the end
unless the end cannot be attained without them, as life cannot be
preserved without food, or cannot be so well attained, as in the case of a

---

1. Chap. 15.
2. Chap. 23.

journey without a horse. But the end of the divine will, by which things were brought into existence, cannot be anything else than God's goodness, as was shown in Book I,[3] and the divine goodness does not depend on creatures either for its existence, since it necessarily exists in itself, or for its well-being, since it is absolutely perfect in itself. All this has been proved above.[4] Therefore no creature's existence is absolutely necessary; consequently we do not have to hold that some creature has always existed.

Again, nothing that proceeds from the will is absolutely necessary, unless the will happens to be impelled by necessity to will something. But God brings creatures into existence, not by any necessity of His nature, but by His will, as we have proved;[5] nor does He necessarily will creatures to be, as was shown in Book I.[6] Hence it is not absolutely necessary for any creature to be, and therefore it is not necessary that creatures should always have existed.

Furthermore, we showed above that God does not act with an action that is outside Himself;[7] that is, no action of His issues forth from Him to terminate in a creature, in the way that heat issues from fire and terminates in wood. God's willing is His acting, and things are as God wishes them to be. But God is under no necessity to will that a creature should have existed always, since He does not have to will that a creature should exist at all, as was shown in Book I.[8] Therefore it is not necessary for any creature to have existed always.

Besides, a thing does not proceed necessarily from a voluntary agent except by reason of some obligation. But, as was shown above,[9] God is under no obligation to produce creatures, if the universal production of creatures is considered absolutely. Therefore God does not produce a creature out of necessity. Consequently, although God is eternal, it is not necessary for Him to have produced any creature from eternity.

Moreover, we proved above[10] that absolute necessity in created things results, not from a relationship to the first principle that is necessarily self-existent, but from a relationship to other causes that are not necessarily self-existent. But the necessity that stems from a relationship to something that is not necessarily self-existent, does not require a thing to have existed always. Thus, if something runs, it must be in motion; but

---

3. Chaps. 75-81.
4. Book I, chaps. 13, 28.
5. Chap. 23.
6. Chap. 81.
7. Chaps. 9, 23.
8. Chap. 81.
9. Chap. 28.
10. Chap. 30.

it need not have been always in motion, because the act of running itself is not absolutely necessary. Consequently, there is no cogent reason why creatures should have existed always.

Chapter 32

Arguments Of Those Who Seek To Prove The Eternity Of The World From God's Standpoint

However, since many have held that the world has existed always and of necessity, and have endeavored to demonstrate their position, we must present their arguments, with a view to showing that they do not necessarily lead to the conclusion that the world is eternal. We shall set forth, first, arguments derived from God's standpoint; secondly, those that are derived from the point of view of creatures;[11] thirdly, those that are derived from the manner in which things were made, according to the contention that they began to exist anew.[12]

To prove the eternity of the world from God's standpoint, the following arguments are alleged.[13]

1. Every agent that is not always in action, is moved either directly or indirectly: directly, as when a fire, which was not always burning, begins to burn, either because it is freshly kindled or is just now carried over and placed next to some fuel; indirectly, in the way that an agent begins to move an animal by some new motion that affects it, either from within, as when an animal awakening after the digestive process is finished, begins to roam around, or from without, as when some fresh incentives entice the animal to undertake some new action. But God is not moved either directly or indirectly, as was proved in Book I.[14] Therefore God always acts in the same way, and it is by His action that created things are brought into being. Accordingly, creatures have always existed.

2. Again, an effect proceeds from its efficient cause by the action of that cause. But God's action is eternal; otherwise He would become an actual agent after being a potential agent, and He would have to be actualized by some prior agent; which is impossible. Therefore the things created by God have existed from eternity.

---

11. Chap. 33.
12. Chap. 34.
13. These arguments are refuted in chap. 35.
14. Chap. 13.

3. Furthermore, given a sufficient cause, its effect must necessarily follow. For if, given the cause, the effect does not follow with necessity, it would be possible, given the cause, that the effect could exist and yet not exist; but what is possible needs something to reduce it to act. Hence some cause will have to be found to make the effect actual; and so the first cause was not sufficient. God, however, is the sufficient cause of the production of creatures; otherwise He would not be a cause in reality, but would rather be a cause in potentiality, since He would become a cause by the addition of something; and that is clearly impossible. Accordingly, since God has existed from eternity it seems that the creatures too, must have existed from eternity.

4. Likewise, a voluntary agent does not delay in carrying out his purpose of making something unless he is awaiting some future event that has not yet taken place. And this sort of thing sometimes occurs within the agent himself, as when full power for acting or the removal of some obstacle impeding such power is awaited. Sometimes, too, it is outside the agent, as when someone in whose presence the action is to be performed is expected, or at least when a suitable time that has not yet arrived is awaited. For if the will is fully equipped, its power is straightway carried into execution, unless it labors under some defect; thus the movement of a limb immediately follows the will's command, if no defect is found in the motive power that is to carry out the movement. Hence it is clear that, when one wishes to do something and it is not done at once, the failure must be ascribed either to a defect in power, the removal of which is looked for, or the fact that the will is not fully equipped to perform the action. And I assert that the will is fully equipped when one wills absolutely to do something no matter what the circumstances may be; whereas the will is insufficiently equipped when a person wills to do something, not absolutely, but on some condition that is not yet verified, or unless a present obstacle is removed. Evidently, however, that which G od now wills to exist, He has eternally willed to exist, for no new movement of the will can occur in Him. Nor could any defect or impediment obstruct His power; and nothing else could be awaited as an incentive for the universal production of creatures, since apart from God nothing is uncreated, as was proved above.[15] Therefore the conclusion that God has brought creatures into existence from eternity seems imperative.

5. Besides, an intellectual agent does not choose one thing in preference to another unless the one is more excellent than the other. But where there is no difference, there cannot be any superiority in excellence. Hence, where no difference is discernible, no choice will be made

---

15. Chap. 15.

of one thing rather than of another. This is the reason why no action will proceed from an agent equally indifferent to either of two alternatives, just as no action proceeds from matter; for potentiality of this sort is like the potentiality of matter. But there can be no difference between non-being and non-being. Hence one non-being is not preferable to another.

Now beyond the entire universe of creatures there is nothing except the eternity of God. But no difference in periods of time can be designated in nothingness; hence we cannot say that a thing ought to be made at one moment rather than at another. Likewise, such a difference in moments cannot be designated in eternity, the whole of which is uniform and simple, as was shown in Book I.[16]   The consequence is, therefore, that God's will is equally disposed to the production of creatures throughout all eternity. Accordingly, it is His will either that creatures should never be produced within His eternity, or that they should always have been produced. Obviously, however, it is not His will that creatures should never have been produced throughout His eternity, since creatures evidently came into being at the behest of His will. The conclusion, therefore, that the created world has always existed, is apparently necessary.

6. Again, things that are directed to an end derive their necessity from that end; this is especially true of things that are done voluntarily. Hence, as long as the end remains the same, things directed to the end preserve the same dispositions or are produced in the same way, unless a new relation arises between them and the end. But the end of creatures issuing forth from the divine will is the divine goodness, which alone can be the end of the divine will. Since, therefore, the divine goodness remains the same throughout all eternity, both in itself and in reference to the divine will, creatures are apparently brought into being by the divine will in the same manner throughout all eternity. For if they are assumed to be quite without existence prior to some definite time at which they are supposed to have begun their existence, they cannot be said to have acquired any new relation to their end.

7. Moreover, since the divine goodness is supremely perfect, all things are said to have come forth from God by reason of His goodness, not in the sense that any advantage might accrue to Him from creatures, but in the sense that goodness has the tendency to communicate itself so far as possible; goodness is manifested by such self-communication. Since all things share in the goodness of God so far as they have being, the more permanent they are, the more they share in God's goodness[17]; that is why the perpetual existence of a species is said to be divine. But the divine goodness is infinite. Accordingly it has the tendency to commu-

---

16. Chap. 15.
17. Cf. Aristotle, *De anima,* II, 4 (415b 5).

nicate itself in an infinite way, and not merely at a particular time. Therefore the divine goodness seems to be a reason requiring the eternal existence of some creatures.

Such, then, are the arguments drawn from the side of God; they seemingly lead to the conclusion that creatures have existed forever.

Chapter 33

Arguments of Those Who Seek to Prove the Eternity of the World from a Consideration of Creatures

There are also other arguments, derived from the viewpoint of creatures, that seem to indicate the same conclusion.[18]

1. Things that have no potentiality to non-existence cannot be without existence. But among creatures there are some that have no potentiality to non-existence. For only creatures that possess matter subject to contrariety can have potentiality to non-existence, since potentiality to existence and non-existence is potentiality to privation and form, whose subject is matter; and privation is always connected with the contrary form, since matter cannot exist without any form at all. But there are certain creatures that have no matter subject to contrariety, either because they are entirely devoid of matter, such as intellectual substances, as will be shown later on,[19] or because they have no contrary, such as heavenly bodies, as is proved by their movement, which has no contrary. Hence it is possible for certain creatures not to exist. Of necessity, therefore, they must exist always.

2. Likewise, every thing endures in being in proportion to its power of being, except by accident, as in those things that are destroyed by violence. But some creatures are endowed with power to exist, not for a limited time, but forever. Such are heavenly bodies and intellectual substances; they are incorruptible because they have no contrary. Consequently everlasting existence is their due; on the other hand, that which begins to exist, does not always exist. In the case of incorruptible things, therefore, a beginning of existence is out of the question.

3. Furthermore, whenever anything begins to be moved for the first time, either the mover or the thing moved or both must behave otherwise now, when there is movement, than previously, when there was no movement. For, according as a mover actually moves, it has a certain ref-

---

18. These arguments are answered in chap. 36.
19. Chap. 50.

erence or relation to the thing that is moved. However, a new relation does not arise without some change in both or at least one of the extremes. But a thing that is in a different state now than formerly, is moved. Therefore, prior to newly initiated movement, another movement, either in the movable thing or in the mover, has to precede. Accordingly, every movement must either be eternal or must have some other movement preceding it. Hence there was always movement, and consequently there were always movable things. And so creatures have always existed, since God Himself is absolutely immovable, as was proved in Book I.[20]

4. Besides every agent that generates its like intends to preserve perpetual existence in the species, since existence cannot be perpetually preserved in the individual. But natural desire cannot be futile. Therefore the species of things capable of generation must be perpetual.

5. Further, if time is everlasting, motion, too, must be everlasting, since time is the measure of motion.[21] Consequently movable things also must be everlasting, since movement is the act of the movable.[22] But time must be everlasting, for time is inconceivable without a *now*, just as a line is inconceivable without a point. But *now* is always the end of the past and the beginning of the future;[23] this is the very definition of *now*. And so any given *now* has time preceding and following it. And thus no *now* can be either first or last. Consequently movable things, which are created substances, must exist from eternity.

6. Again, we must either affirm or deny. Therefore, if the affirmation of a thing is implied in its negation, that thing must always exist. Such a thing is time. For if we say that time did not always exist, we assert that it did not exist before it existed; similarly, if time is not to exist always in the future, its non-existence must come after its existence. But there can be no before or after in duration unless there is time, since time is the measure of before and after.[24] And thus time would have to exist before it began, and continue into the future after it ceased. Accordingly time must be eternal. Now time is an accident, and an accident cannot exist without a subject. Its subject, however, is not God, who is above time; for He is absolutely immutable, as was proved in Book I.[25] Consequently some created substance is eternal.

7. Moreover, many propositions are of such a kind that he who denies them must affirm them. For instance, whoever denies that there is truth, supposes that truth exists, for he supposes that the negative

20. Chap. 13.
21. Cf. Aristotle, *Physics,* IV, 11 (219b 1).
22. Cf. Aristotle, *Physics,* III, 2 (202a 7).
23. Cf. Aristotle, *Physics,* IV, 13 (222b 1).
24. Cf. Aristotle, *Physics,* IV, 11 (219b 1).
25. Chap. 13.

proposition he utters is true. The case is similar with him who denies the principle that contradictories are not simultaneous, for by denying that he asserts that the denial he voices is true and that the opposite affirmation is false, and thus that both are not verified about the same thing. Therefore, if a thing whose denial entails its affirmation must exist always, as we have just proved, the aforesaid propositions and all propositions derived from them are eternal. But these propositions are not God. Therefore something besides God must be eternal.

These arguments, then, and others like them can be adduced on the part of creatures to prove that creatures have always existed.

Chapter 34

Arguments to Prove the Eternity of the World From the Viewpoint of the Creative Action

Other reasons can be alleged on the part of the creative action itself to prove the same conclusion.[26]

1. What is commonly asserted by all, cannot be wholly false. For a false opinion betrays a certain weakness of intellect, just as a false judgment about a proper object of sense results from a weakness in the sense. But defects are accidental, because they are outside the intention of nature. What is accidental, however, cannot exist always and in all beings; thus the judgment which every taste registers about flavors cannot be false. Accordingly a judgment that is pronounced by all men about a truth cannot be erroneous. Now it is the common opinion of all philosophers that nothing is made from nothing.[27] Consequently it must be true. Therefore, if anything is made, it must be made from something. And if the latter thing is also made, it too, must be made from something else. But this process cannot go on indefinitely, for then no generation would ever be completed, since an infinite number of stages cannot be traversed. Hence we must come to some first thing that was not made. But every being that has not existed forever must have been made. Therefore that being from which all other things were originally made must be eternal. But this being is not God, because He cannot be the matter of anything, as was proved in Book I.[28] Consequently something outside of God must be eternal, namely, prime matter.

---

26. These arguments are refuted in chap. 37.
27. Cf. Aristotle, *Physics,* I, 4 (187a 28).
28. Chap. 17.

2. Moreover, if a thing is not in the same state now as before, it must have undergone some change, for this is precisely what we mean by mutation: that a thing is not in the same state now as before. But everything that begins to exist anew, is not in the same state now as before; and the reason is, that some motion or change has taken place. All motion or change occurs in some subject, for motion is the act of a movable thing.[29] Since motion precedes that which is brought about by it, seeing that motion terminates in such a thing, some movable subject must exist prior to anything that is made. And as an infinite regression along this line is impossible, we must come to some first subject that never began but always existed.

3. Besides, everything that begins to exist anew, had a possible existence before it actually existed; otherwise it was impossible for it to exist and necessary for it not to exist. Thus it would forever have been a non-being and would never have begun to be. But that which has the possibility of existing is a subject that is potentially a being. Therefore, before anything begins to exist anew, a subject that is potentially a being must pre-exist. And since we cannot go back in this way indefinitely, we must suppose some primary subject that did not begin to exist anew.

4. Again, no permanent substance exists while it is being made. For it is made that it may exist, hence if it were already existing, it would not still have to be made. But while it is being made, there must be some subject that is worked upon; for, since the process of making is an accident, it requires a subject. Accordingly everything that is made has some pre-existing subject. And since this cannot go on indefinitely, the conclusion follows that the first subject was not made, but is eternal. From this the further conclusion follows that something besides God is eternal, for He Himself cannot be the subject of making or of movement.

These, then, are the arguments to which some thinkers give their adherence, as though they were demonstrations vindicating their contention that created things must have existed forever. In this they contradict the Catholic faith, which insists that nothing outside God has existed forever, and that all things, except the one eternal God, have had a beginning of their existence.

---

29. Cf. Aristotle, *Physics,* III, 2 (202a 7).

Chapter 35

Solution of the Arguments Alleged Above, and First of Those Derived from the Standpoint of God

We have now to show that the foregoing arguments do not necessarily conclude. We shall consider, first, those that are alleged on the part of the agent.[30]

Even if the effects produced by God begin to exist anew, He Himself need not be moved either directly or indirectly, as the *first* objection argued. Newness of effect can, indeed, indicate a change in the agent to the extent that it demonstrates newness of action; a new action cannot be performed by an agent unless the latter is moved in some way, at least from inactivity to action. Yet newness of a divine effect does not demonstrate newness of action in God, since His action is His essence, as was brought out above.[31] And therefore newness of effort cannot demonstrate such change when God is the agent.

Nevertheless it does not follow that, if the action of the First Agent is eternal, His effect must be eternal, as the *second* argument inferred. We showed above that God acts voluntarily in producing things.[32] However, He does not act in such a manner that some other, intermediate action is elicited by Him, in the way that in us the action of our motive power intervenes between the act of the will and the effect, as we have shown previously;[33] God's act of understanding and willing is necessarily identical with His act of making. An effect issues from the intellect and will according to the direction of the intellect and the command of the will. Just as any other condition of the thing made is determined by the intellect, so a time is set for it; art not only determines that a thing should be of a definite quality, but assigns a time for it. For example, a physician prescribes that a medicine should be given at this or that time; if his act of will were in itself powerful enough to produce the effect, the effect would follow in due time from the previous decision, without any new action on his part. Hence there is nothing to prevent us from saying that God's action existed from eternity, although its effect was not produced from eternity but occurred at the time eternally appointed for it.

---

30. Cf. chap. 32.
31. Chap. 9.
32. Chap. 23.
33. Chap. 9.

This consideration also makes it clear that, even though God is the sufficient cause producing the existence of things, we do not have to conclude that His effect is eternal just because He Himself exists eternally, as the *third* argument contended. Given a sufficient cause, its effect does, indeed, follow. This is not true, however, of an effect that is foreign to the cause, such as would result from an insufficiency in the cause, as if, for example, a hot body failed to emit heat. The proper effect of the will is the production of that which the will decides; if something else than what the will decrees were to result, the effect would not be proper to the cause but would be alien to it. But, to repeat what we have said, just as the will determines that a thing should be of a definite nature, so it wills that the thing should exist at a particular time. In order that the will may be a sufficient cause, therefore, the effect need not exist when the will itself exists, but only at the time which the will appoints for it. The case is different with things proceeding from a cause that acts naturally, because the action of nature is determined by the nature itself; given the existence of the cause, the effect must follow. The will, however, acts in a way that is governed, not by its existence, but by its intention. And therefore, just as the effect of a natural agent is determined by the agent's existence, if the latter is a sufficient cause, so the effect of a voluntary agent is produced in accord with the agent's purpose.

All this makes it manifest that the effect of the divine will is not unduly retarded, as the *fourth* argument suggested, even though it did not always exist, notwithstanding the fact that it was willed. Not only the existence of the effect, but the time of its existence, is subject to the divine will. Therefore the thing willed, that is, the existence of a creature at a definite time, is not retarded, because the creature began to exist at the moment appointed by God from eternity.

We cannot admit any diversity of parts in some sort of duration prior to the beginning of all creation, as was supposed in the *fifth* argument. For nothingness has neither measure nor duration. And God's duration, which is eternity, has no parts, but is absolutely simple, without before and after, since God is immovable, as was explained in Book I.[34] Hence there is no question of comparing the beginning of all creation with diverse parts designated in some pre-existing measure, with which the beginning of creatures might be in agreement or disagreement, so that an agent would have to have a reason for bringing creatures into existence at this particular instant of that duration rather than at some preceding or subsequent instant. Such a reason would indeed be required if some duration divisible into parts existed outside the universe of created beings, as is the case with particular agents that produce their effect in time, yet

---

34. Chap. 15.

do not produce time itself. But God brought creatures and time into being together. Hence the question to consider is not, why He produced them now rather than earlier, but only why He did not endow them with eternal existence. Comparison with place will clarify this point. Particular bodies are produced in a definite place, as well as at a definite time. And because time and place, by which they are enveloped, are extrinsic to them, there must be a reason why they are produced in this place and at this time rather than another. But there is no reason for inquiring why the whole of heaven was located here and not there, for beyond it there is no place, and the entire place for all things was produced along with it. In their quest of such a reason, some fell into the error of attributing infinity to bodies. Similarly, outside the totality of creation there is no time, since time was produced simultaneously with the universe; hence we need not go into the reason why it exists now and not earlier, and so we will not be led to concede the infinity of time. We have only to ask why it did not exist always, or why it came into being after non-being, or why it had a beginning at all.

To pursue this inquiry the *sixth* argument was proposed; it is based on a consideration of the end, which alone can induce necessity in things that are done voluntarily. The end of the divine will cannot be anything else than God's goodness. However, God does not act in order to bring this end into being, in the way that a craftsman acts in constructing his handiwork; for God's goodness is eternal and unchangeable, so that nothing can be added to it. Nor can God be said to act for His own self-improvement. Again, He does not act to acquire this end for Himself, as a king engages in warfare to gain possession of a city, for God is His own goodness. It remains, consequently, that He acts for an end in the sense that He produces an effect which is to share in the end. In this production of things for the sake of an end, the uniform relation of end to agent is not to be regarded as an argument in favor of the eternity of His work; we should rather focus our attention on the relation of the end to the effect that is produced for the end. Thus we perceive that the effect is produced in such a way as to be more fittingly directed to the end. Hence the fact that the end is uniformly related to the agent does not justify us in concluding that the effect is eternal.

Lastly, it is not necessary that the divine effect should have existed always, for the reason that it would thus be more suitably directed to the end, as the *seventh* argument seemed to imply. Indeed, it is more suitably directed to the end for the very reason that it did not exist always. For every agent that produces an effect designed to participate in the agent's own form aims at reproducing its own likeness in the effect. Fittingly, therefore, God willed to produce creatures for participation in His goodness; by resembling the divine goodness they would represent it. Such a representation cannot reach the level of equality, in the way that a univocal effect represents its cause, so that effects produced by infinite goodness would have to be eternal; no, God is represented by creatures as the transcendent is represented by that which is surpassed. But the transcendence of the divine goodness over creation is best brought out by the fact that creatures have not always existed. This fact clearly shows that all things outside of God have Him as the author of their being, and that His power is under no necessity to produce such effects, as nature is with regard to natural effects; and consequently that He is a voluntary and intelligent agent. Some thinkers have entertained views opposed to these truths, because they assume the eternity of creatures.

Accordingly there is nothing on the part of the agent that compels us to hold the eternity of creation.

Chapter 36

Solution of the Arguments Alleged on the Part of the Things Produced

Likewise there is nothing on the part of creatures that necessarily induces us to assert their eternity.[35]

The necessity of existing which is found in creatures and is the basis of the *first* argument for this position, is a necessity of order, as was shown previously.[36] But the necessity of order does not force the thing that is subject to such necessity to have existed forever, as was pointed out above.[37] Although the substance of heaven necessarily exists, since it lacks potentiality to non-existence, this necessity is consequent on its substance. Therefore, once heaven's substance has been brought into being, this necessity entails the impossibility of not existing; but if the production of its substance is taken into consideration, such necessity does not obviate heaven's non-existence.

Similarly, the power of existing always, from which the *second* argument was drawn, presupposes the production of the substance. Therefore, since the production of the substance of heaven is in question, such power cannot be a conclusive argument for the eternity of that substance.

---

35. Cf. chap. 33.
36. Chap. 30.
37. Chap. 31.

The objection next adduced does not compel us to acknowledge the eternity of motion. It is plain from the preceding chapter that without any change in Himself, God the agent can accomplish something new, something that is not eternal. But if something can be done by Him anew, it is clear that something can also be put in motion by Him anew, for newness of motion follows on the decision of the eternal will that motion is not to be eternal.

Likewise, the tendency of natural agents to perpetuate their species, which was the source of the *fourth* argument, presupposes that natural agents have already been produced. Hence this argument is out of place except with regard to natural things that have already been brought into being; it is irrelevant when there is question of the production of these things. Whether a process of generation that will go on forever has to be admitted, will be examined later.[38]

Also the *fifth* argument, based on time, presupposes rather than proves the eternity of motion. According to Aristotle's teaching, the *before* and *after* and continuity of time are dependent on the *before* and *after* and continuity of motion;[39] hence the same instant is clearly the beginning of the future and the end of the past, because some assignable point in motion is the beginning and end of the various segments of motion. Not every instant, therefore, will have to be of this kind, unless every assignable point in time is taken to be midway between before and after in motion - which is to suppose that motion is eternal. But if we suppose that motion is not eternal, we can say that the first instant of time is the beginning of the future without being the end of any past. Nor is the contention that some *now* is a beginning and not an end incompatible with the succession of time, on the ground that a line, in which a point is placed as a beginning and not an end, is stationary and not transitory; for even in a particular movement, which of course is not stationary but transitory, some point can be designated as only the beginning without being the end of movement. Otherwise all motion would be perpetual, which is impossible.

In the hypothesis that time began, the assertion that it did not exist before it existed does not compel us to admit that the very supposition of its non-existence implies its existence, as the *sixth* argument inferred. For the *before* we speak of as preceding time does not imply any flight of time in reality, but only in our imagination. When we say that time exists after not existing, we mean that no time was flowing prior to this designated *now*. In the same way, when we say that there is nothing above heaven, we do not mean that there is some place beyond heaven which may be said to be *above* in relation to it, but that there is no place

---

38. Book IV, chap. 97.
39. *Physics,* IV, 11 (219a 17).

higher than heaven. In both cases the imagination can add a certain dimension to the existing thing. However, just as such addition is no reason for attributing infinite quantity to a body, as is indicated in Book III of the *Physics*,[40] so neither is it a reason for admitting that time is eternal.

The truth of the propositions which even the one who denies the propositions must concede, as maintained in the *seventh* argument, involves the necessity of the order which exists between predicate and subject. Such necessity does not require a thing to exist forever, except, indeed, the divine intellect, in which all truth is rooted, as was shown in Book I.[41]

Clearly, therefore, the arguments derived from creatures do not compel us to assert the eternity of the world.

Chapter 37

Solution of the Arguments Drawn from the Creative Action

Our remaining task is to show that none of the arguments derived from the viewpoint of the production of things imposes the same conclusion.[42]

The position common to philosophers who contend that nothing is made from nothing, which is the starting point of the *first* argument, is true with regard to the kind of making they had in mind. Since all our knowledge begins with sense perception, which has to do with individual things, human speculation mounted from particular to universal considerations. Hence they who investigated the origin of things considered only particular instances of production, and inquired how this fire or this stone came into being. And so they who came first, limiting their consideration of the production of things to an excessively superficial examination, maintained that a thing is made merely according to certain accidental modifications, such as rarity, density, and the like, and asserted in consequence that "to be made" was nothing but "to be altered"; that is, they thought that everything was made from something actually existing. But their successors, who entered more deeply into the process by which things are made, advanced to a consideration of the production of things

---

40. Aristotle, *Physics,* III, 6 (206b 20).
41. Chap. 62.
42. Cf. chap. 34.

in terms of their substance; they taught that a thing need not be made from actually existing being, except in accidentals, and that in its essentials it is made from potential being. Yet this sort of production, which is that of a being made from some other being, is the making of a particular being, one that is made inasmuch as it is this being, such as a man or a fire, but not universally, inasmuch as it simply is; for there existed previously a being that was transformed into this being. Penetrating, then, still more deeply into the origin of things, they considered at last the procession of all created being from a single first cause, as is clear from the arguments listed above to prove this truth.[43] In this procession of all being from God, it is impossible for anything to be made from something else previously existing; otherwise it would not be the making of created being in its entirety.

This kind of making eluded the early naturalists, who shared the common opinion that nothing is made from nothing. Or, if any of them did arrive at the notion, they did not think that the term "making" properly expressed it, since the word "making" implies motion or change, whereas in this origin of all being from one first being, the transformation of one being into another is inconceivable, as we have shown.[44] For this reason, the investigation of this kind of origin of things is the concern, not of the philosopher of nature, but of the metaphysician, who studies universal being, including such things as are dissociated from motion. By virtue of a certain similarity, however, we transfer the term "making" even to such origin, and so we say that those things are made whose essence or nature takes its origin from other beings.

Consequently it is evident that the *second* argument, which is based on the nature of motion, does not lead to a necessary conclusion. Creation cannot be called change except metaphorically, so far as a created thing is regarded as having existence after non-existence. In this sense one thing is said to issue from another even when no change from one thing into another takes place, for the sole reason that one of them succeeds the other, as day comes forth from night. Nor can the nature of motion that is here alleged be of any avail, for what does not exist at all is not in any particular state; hence we cannot conclude that, once a thing begins to exist, it is in a different state now from what it was before.

Therefore it is also plain that no passive potentiality has to precede the existence of all created being, as the *third* argument attempted to prove. Such a necessity does, indeed, obtain in things that originally come into being by way of motion, seeing that motion is the act of a being existing in potentiality. Before a created being existed, its existence was possible, because of the power of the agent that initially endowed it with exis-

43. Cf. chap. 16.
44. Chap. 17.

tence, or by reason of the compatibility of the terms, in which no contradiction is found. Possibility in this latter sense involves no reference to potentiality, as the Philosopher asserts in Book V of his *Metaphysics.*[45] For the predicate "to be," is not incompatible with the subject, "world" or "man," as "commensurate" [predicated of the side of a square] is incompatible with "diagonal." And thus it follows that the existence of the world or of a man is not impossible, and consequently that before they existed their existence was possible, even in the absence of potentiality. But things that come into being by way of motion must previously be possible by reason of some passive potentiality; such are the things the Philosopher has in mind when he employs this argument in Book VII of the *Metaphysics.*[46]

This also explains why the *fourth* argument fails to prove its point. For in things that are made by way of motion, *to be made* and *to be* are not simultaneous, since their production involves succession. But in things that are not made by way of motion, their making does not precede their existence.

Therefore it is quite evident that nothing prevents us from maintaining that the world has not always existed. And this is what the Catholic faith teaches: "In the beginning God created heaven and earth" (Genesis 1:1); and in Proverbs 8:22 it is said of God: "Before He made anything from the beginning," etc.

Chapter 38

Arguments by Which Some Endeavor to Prove That the World is not Eternal

We come now to a group of arguments brought forward by certain thinkers to prove that the world has not always existed; they are drawn from the following considerations.

1. That God is the cause of all things, has been demonstrated. But a cause must precede in duration the things that are produced by its action.

2. Again, since all being has been created by God, it cannot be said to have been made from some being, and so we conclude that it was made from nothing. Consequently its existence must be subsequent to non-existence.

---

45. Aristotle, *Metaph.,* V, 12 (1019b 34).
46. Aristotle, *Metaph.,* VII, 7 (1032a 15).

3. Another reason is that an infinite series cannot be crossed. But if the world has existed forever, an infinite series would by now have been crossed. For the past has been traversed, and if the world has always existed, an infinite number of days or of solar revolutions have gone by.

4. A further consequence is that an addition is made to the infinite, for new additions are daily being made to the days or solar revolutions which have passed.

5. Still another consequence is that we are proceeding to infinity in efficient causes, in the hypothesis that generation has always taken place. This has to be admitted if the world has existed forever, for the father is the cause to his son, and another man is the cause of the father, and so on, back into the endless past.

6. Furthermore, it will follow that an infinite multitude exists, namely, the immortal souls of an infinite number of men dead and gone.

Since these arguments do not conclude with strict necessity, although they are not entirely devoid of probability, it is enough to touch on them briefly, so that the Catholic faith may not seem to rest on inept reasonings rather than on the unshakable basis of God's teaching. And so we deem it suitable to show how such arguments are met by those who have maintained the eternity of the world.

The *first* contention, that an agent necessarily precedes the effect wrought by its action, is true of causes which produce something by way of motion, because the effect is not achieved until the motion has come to a stop, whereas the agent must exist even when the motion begins. But in causes that act instantaneously the same conclusion need not follow. Thus, as soon as the sun reaches the point of the east, it at once lights up our hemisphere.

Also the *second* argument is invalid. For if the proposition, "Something is made from something," is not granted, the contradictory to be asserted is, "Something is not made from something," and not, "Something is made from nothing," except in the latter sense. We cannot conclude from this that a thing comes into existence subsequent to non-existence.

Again, the *third* argument lacks cogency. For, even though the infinite does not exist all at once if it is actual, it can exist successively; understood in this sense, any infinite is finite. Therefore, since each of the preceding revolutions was finite, it could make a full turn. But if they are all viewed as existing simultaneously, on the supposition that the world had always existed, a first revolution cannot be assigned. And so there could be no transition to the present, for transition always requires two extremes.

The *fourth* argument, too, is weak. There is nothing to prevent an addition to the infinite on the side on which it is finite. On the supposition that time is eternal, it must be infinite on the part of what went before, but finite on the part of what came after, for the present marks the end of the past.

The *fifth* objection is likewise devoid of cogency. According to philosophers, infinite procession is impossible when there is question of efficient causes that act together at the same time, because the effect would have to depend on an infinite number of co-existing actions. But according to those who propose the theory of an endless series of generations, infinite procession is not impossible in the case of causes that do not act simultaneously. Here the infinity is accidental to the causes; whether the father of Socrates is another man's son or not, affects him only accidentally. But when a stick moves a stone, its movement by the hand that holds it is not accidental; the stick moves only to the extent that it is moved.

The *sixth* objection, about souls, is more difficult. Yet the argument is not of much utility, as it supposes many things.[47] Some advocates of the eternity of the world have contended that human souls do not survive the body. Some have maintained that nothing remains of all the souls except a separate intellect: the agent intellect, in the opinion of some, or also the possible intellect, in the view of others. Some have imagined a rotation in souls, saying that the same souls return to bodies after several centuries. And some hold that it is not impossible for certain things to be actually infinite in number, if there is no order among them.

A more effective procedure can be adopted in this matter if we start with the purpose of the divine will, as was suggested above.[48] The end of the divine will in the production of things is the goodness of God as manifested by His effects. But the divine power and goodness are best made known by the fact that things other than God Himself have not existed forever. The very fact that such things have not always existed shows clearly that they have their existence from Him. It also shows that God does not act by necessity of His nature, and that His power of acting is infinite. To manifest the divine goodness, therefore, it was supremely fitting that God should assign to created things a beginning of their duration.

---

47. Cf. Book II, chap. 81, in the response to the third argument. The objection takes for granted that man has always inhabited the earth and that an infinite multitude of spiritual substances is impossible. St. Thomas points out that the arguments used by Aristotle to prove the impossibility of an actual infinity of material bodies do not touch the question of an infinite multitude of immaterial beings. He adds that this difficulty cannot disturb those who profess the Catholic faith, since Catholics do not admit the eternity of the world.
48. Chap. 35.

Reflections such as these enable us to avoid various errors into which pagan philosophers fell. Some of them asserted the eternity of the world. Others held that the matter of the world is eternal, and that the world began to be fashioned from it at a certain moment, either by chance, or at the direction of some intellect, or else by attraction and repulsion. But all of them take it for granted that something besides God is eternal; and this is incompatible with the Catholic faith.

St. Thomas Aquinas

On the Power of God
(De Potentia Dei)

Question 3, Article 17

Whether the World Has Existed Forever

The seventeenth inquiry is whether the world has existed forever. The answer would seemingly be affirmative, for the following reasons.

1. A thing always achieves what is proper to it. But, as Dionysius points out, it is proper to God to summon beings which exist to a share in the divine goodness;[1] and this is accomplished by producing creatures. Since, therefore, the divine goodness has always existed, God has apparently always brought creatures into being; and so we may conclude that the world has existed forever.

2. God does not withhold from any creature what is within its capacity in accordance with its nature. But there are some creatures with a nature capable of having always existed, as, for example, heaven. Therefore heaven seems to have received the gift of eternal existence. And if we grant that heaven existed, we ought also to grant that other creatures existed, as the Philosopher shows in Book II of *De Caelo et mundo.*[2] Therefore it seems that the world has always existed. Proof of the minor: Whatever is incorruptible has the power of existing forever; for if it had the power of existing for a limited time only, it could not exist forever, and hence it would not be incorruptible. However, heaven is incorruptible. Therefore it is capable by nature of existing always.

3. But someone may object that heaven is not absolutely incorruptible, for it would drop back into nothingness if it were not sustained in being by God's power. To this objection the reply may be made that a thing is not to be judged possible or contingent on the ground that its destruction results from the removal of some consequent fact; thus, although man is necessarily an animal, his destruction ensues upon the removal of the consequent fact that man is a substance. Therefore we ought not to conclude that heaven is corruptible merely because it would cease to exist on the supposition that God were to withdraw His sustaining influence from creatures.

---

1. *De caelesti hierarchia,* cap. 4 (PG 3, 177).
2. Aristotle, *De caelo,* II, 3 (286a 10-b 9).

4. As Avicenna proves in his *Metaphysics*, every effect is necessary in relationship to its cause.[3] For if, given the cause, the effect does not necessarily follow, it will be possible for the effect to follow or not to follow even when the cause is present. But what is in potentiality is not reduced to act except by something actual; hence, in addition to the aforesaid cause, some other cause will be needed to render the effect actual from the state of potentiality in which, even in the presence of the cause, it had the possibility of existing or of not existing. From this we can gather that when there is a sufficient cause, the effect must necessarily follow. But God is a sufficient cause of the world. Therefore, since God has always existed, the world, too, has always existed.

5. Whatever precedes time is eternal, aeviternity, however, does not precede time, but began along with time. But the world did precede time, since it was created in the first instant of time, and that first instant certainly preceded time; for we are told in Genesis 1:1: "In the beginning God created heaven and earth," that is, in the beginning of time. Therefore the world existed from eternity.

6. As long as anything remains one and the same, it always produces the same effect, unless it is hindered. But God always remains one and the same, as we read in Psalm 101:28: "Thou art always the self-same." Accordingly, since He cannot be hindered in His action, because His power is infinite, we should conclude that He always produces the same effect. And therefore, since He produced the world at some time, He apparently has always produced it, from eternity.

7. Just as man necessarily desires his own beatitude, so God necessarily wills His own goodness and all that pertains to it. But the production of creatures pertains to the divine goodness. Therefore this is something which God necessarily wills; and so, as He has willed His own goodness from eternity, it seems that He has also willed to produce creatures from eternity.

8. The objection may be raised that it does, indeed, pertain to God's goodness to bring creatures into existence, but not to do so from eternity. The answer to this is that the bestowal of a gift early rather than later on, evinces greater generosity, and the generosity of the divine goodness is infinite. Therefore God has seemingly endowed creatures with existence from eternity.

9. Augustine says that a person wills what he does, if he is able to.[4] But from eternity God willed to produce the world; otherwise He would have undergone some change, if a new act of willing to create the world had occurred in Him. Therefore, since no powerlessness can be attributed to Him, He apparently produced the world from eternity.

---

3. *Metaph.*, I, 7 (73 ra).
4. Cf. *Confessions*, VIII, 9 (PL 32, 759).

10. If the world has not always existed, then, before it existed, its existence was either possible or not. If its existence was not possible, then it was impossible, and so its non-existence was necessary; hence it would never have been brought into existence. And if its existence was possible, then there was some potentiality with regard to it, and so there was some substratum or matter, since potentiality requires a subject. But if there was matter, there was also form, since matter cannot be completely devoid of form. Therefore some body composed of matter and form, and consequently the entire universe, was in existence.

11. Whatever becomes actual after it has been possible, is educed from potentiality to actuality. If, therefore, the world was possible before it was brought into existence, we must say that it was educed from potentiality to actuality, and hence that matter preceded it and was eternal; thus we come to the same conclusion as before.

12. Every agent that begins to act anew, passes from potentiality to actuality. But such a process cannot be attributed to God, since He is absolutely changeless. Therefore it seems that He did not begin to act anew, but produced the world from eternity.

13. If a voluntary agent begins to do what he previously willed but did not carry out before, we must suppose that some incentive now induces him to act though it did not induce him previously; such an incentive somehow arouses him to action. But we cannot say that prior to the world there was something beside God which presented Him with a fresh inducement to activity. Since, therefore, He willed to make the world from eternity (otherwise His will would have been subject to some new influence), He apparently did make it from eternity.

14. Nothing moves God's will to act except the divine goodness. But the divine goodness is ever in the same disposition. Therefore God's will also is ever disposed to produce creatures; and so He produced them from eternity.

15. That which is always in its beginning and its end, never begins and never ceases, because a thing exists after it has begun and before it has ceased. But time is always in its beginning and its end, since time is nothing but an instant, which is the end of the past and the beginning of the future. Therefore time never begins and never ends, but exists forever. Consequently, movement and whatever is subject to movement and so also the entire world exist forever; for there is no time apart from movement, or movement apart from things subject to movement, or things subject to movement apart from the world.

16. Here an objector may say that the first instant of time is not the end of the past, and that the last instant is not the beginning of the future. Opposed to this, however, is the fact that the *now* of time is always regarded as flowing, and that in this respect it differs from the *now* of eternity. But whatever flows, proceeds from one point to another.

Therefore every *now* must flow from a previous now to a later one. Consequently any first or last *now* is impossible.

17. Motion follows that which is movable, and time follows motion. But since the first movable body is circular, it has neither beginning nor end, because neither an actual beginning nor an actual end can be designated in a circle. Therefore neither motion nor time has a beginning; and so we have the same conclusion as above.

18. Someone may argue that, although a circular body has no beginning of its magnitude, it has a beginning of its duration. On the contrary, however, the duration of motion follows the measure of magnitude, because, according to the Philosopher, motion and time are in proportion to size.[5] Therefore, if there is no beginning of a circular body's magnitude, there will be no beginning in the magnitude of its motion and time, and consequently there will be no beginning of their duration, since their duration, and especially that of time, is their magnitude.

19. God is the cause of things by His knowledge. But knowledge is predicated in relation to what is knowable. Accordingly, since relatives are by nature simultaneous, and God's knowledge is eternal, it seems that things have been produced by Him from eternity.

20. God precedes the world only in the order of nature, or also in the order of duration. If only in the order of nature, in the way a cause precedes its simultaneous effect, it seems that, since God has existed from eternity, creatures, too, have existed from eternity. And if God precedes the world in duration, we must admit, prior to the duration of the world, some duration which is related to that of the world as *before* and *after*. But duration in which *before* and *after* are discerned, is what we mean by time. Therefore the world was preceded by time, and consequently by motion and movable bodies; and thus we are brought to the same conclusion as before.

21. Augustine states: "I refuse to admit that God was not Lord from eternity."[6] But as long as He was Lord, He had some creature for His subject. Therefore we should not assert that no creature existed from eternity.

22. God could have produced the world before He did produce it, as otherwise He would have been deficient in power. He also knew how to produce it earlier, as otherwise He would have been wanting in knowledge. Seemingly He also willed to create it, as otherwise He would have been subject to envy. Therefore He could hardly have delayed before undertaking to produce creatures.

---

5. Aristotle, *Physics,* IV, 11 (219a 12-20).
6. *De Trinitate,* V, 16 (PI, 42, 922).

23. Any finite perfection can be communicated to a creature. But eternity is something finite; otherwise nothing could extend beyond eternity. Yet Exodus 15:18 has this statement: "The Lord shall reign unto eternity and beyond".[7] Hence we may infer that a creature is capable of eternal existence; and thus it was fitting that the divine goodness should have produced creatures from eternity.

24. Everything that begins to exist, has a measure of its duration. But time cannot have any measure of its duration; it is not measured by eternity, since in that case it would always have existed; nor by aeviternity, as then it would endure forever; nor by time, because nothing is its own measure. Therefore time has no beginning, and the same is true of movable things and of the world.

25. If time had a beginning, it began either in time or in an instant. It did not begin in an instant, because in an instant there is as yet no time. And it did not begin in time, because in that supposition no time would exist before a definite point in time, since before a thing begins to exist it is nothing. Therefore time had no beginning; and again we come to the same conclusion as before.

26. God was the cause of things from eternity; otherwise we should have to say that He was a potential cause before being an actual cause, and so something already in existence would arouse Him from potentiality to act; which is impossible. But there is no cause without an effect. Therefore the world was created by God from eternity.

27. Truth and being are convertible. But many truths are eternal, as, for example, that man is not an ass, that the world was to be, and a host of similar truths. It seems, therefore, that not God alone, but many beings exist from eternity.

28. Against this, someone may argue that all such things are true by reason of the first truth, which is God. On the contrary, however, the truth of this proposition: "The world is to be," differs from the truth of this one: "Man is not an ass," because even though we made the impossible supposition that one of them is false, the other will still be true. But the first truth does not tolerate alteration. Therefore the propositions in question are not true by reason of the first truth.

29. According to the Philosopher, a statement is true or false[8] from the fact that the thing referred to is as it is asserted to be or not. Therefore, if many propositions are true from eternity, the things signified by them have apparently existed from eternity.

---

7. Thus the Vulgate: *Dominus regnabit in aeternum et ultra.* The Douay version has: *The Lord shall rein for ever and ever.*

8. Aristotle, *Categories,* 5 (4b 8).

30. With God, to speak is the same as to accomplish, as is brought out by Psalm 148:5: "He spoke, and they were made." but God's utterance is eternal; otherwise the Son, who is the Father's Word, would not be co-eternal with the Father. Therefore, also, God's act of making is eternal and so the world was made from eternity.

BUT ON THE CONTRARY: 1. In Proverbs 8:24 ff., divine Wisdom is thus represented as speaking: "The depths were not as yet, and I was already conceived; neither had the fountains of waters as yet sprung out; the mountains with their huge bulk had not as yet been established; before the hills I was brought forth. He had not yet made the earth nor the rivers nor the poles of the world." Therefore the poles of the world and the rivers and the earth did not always exist.

2. According to Priscian,[9] the younger in age people are, the keener is their intellectual perception. But such intellectual keenness is not infinite. Therefore the time during which it increased was not infinite, and consequently neither is the world.

3. In Job 14:19 we are told: "With inundation the ground by little and little is washed away." But the earth is not infinite. If, therefore, time had been infinite, the earth would by now have been entirely washed away; and this is evidently false.

4. As is clear, God is prior to the world by nature, in the way that a cause is prior to its effect. But duration and nature are identical in God. Therefore God is prior to the world in duration, and so the world has not existed forever.

I ANSWER THAT we must hold firmly that the world has not always existed, as the Catholic faith teaches. And this truth cannot be effectively attacked by any demonstration based on physics. To make this clear, we must understand, as was brought out in a previous question,[10] that in God's activity we cannot assign any necessity on the part of the material cause, or on the part of the active power of the agent, or on the part of the last end, but only on the part of the form which is the end of the operation; for if a form is presupposed, things must necessarily exist in such a way as to be fit for that form.

And therefore we must speak of the production of one particular creature otherwise than of the creation of the whole universe by God. For, when we speak of the production of some particular creature, we can assign the reason why it is such as it is from a consideration of some other creature, or at least from the order of the universe, to which every creature is subordinated as a part to the form of the whole. But when we

---

9. Priscian was a Neoplatonic philosopher who flourished toward the end of the sixth century A.D. Among the few of his works that has come down to us is the *Commentarium in Tractatum theophrasticum de sensatione.*
10. Art. 16.

speak of the entire universe that is brought into being, we cannot discover any other creature from which we may gather the reason why it is such or such. Accordingly, since a reason for the definite disposition of the universe cannot be discerned either on the part of the divine power, which is infinite, or on the part of the divine goodness, which has no need of created things, the reason for it must be found in the sheer will of the Creator. Hence, if the question should be asked, why the heaven is of such and such a size and not greater, no other reason can be given than that the Creator so willed it.

And because of this consideration, too, as Rabbi Moses says,[11] Sacred Scripture induces men to contemplate the heavenly bodies; their arrangement most clearly shows that all things are subject to the Creator's will and providence. For no reason can be assigned to account for the great distance of this star from that one, or for any other phenomena that may be observed in the disposition of the heavens, except the design of God's wisdom. Thus we are exhorted in Isiah 40:26: "Lift up your eyes on high and see who hath created these things."

This position is not overturned by the contention that a particular quantity is governed by the nature of heaven or of heavenly bodies, on the ground that all existing things have a definite quantity by nature. For, just as the divine power is not restricted to one quantity rather than to another, so it is not restricted to a nature requiring a particular quantity rather than to a nature requiring a different quantity. And so the same question recurs with respect to nature as to quantity, even though we should grant that the nature of heaven is not wholly indifferent to quantity, or that it has no capacity for any other than its present quantity.

But the same cannot be said of time or of the duration of time. For time, like place, is extraneous to a thing. Consequently even heaven, which has no capacity for a different quantity or a different accident intrinsically inhering, has such capacity with regard to place and position, since it has local motion; and also with regard to time, since time ever succeeds time, just as there is succession in movement and locality [ubi]. Hence neither time nor locality can be said to result from the nature of heaven, as was stipulated in the case of quantity. Thus it is clear that the prefixing of a definite quantity of duration for the universe, as also of a definite quantity of dimension, depends on the mere will of God. Accordingly we cannot arrive at any necessary conclusion about the duration of the universe, so as to be able to prove demonstratively that the world has existed forever.

---

11. Moses Maimonides, *Guide for the Perplexed,* II, 19 (p. 187 f).

Some thinkers, however, inevitably fell into error concerning the beginning of the world, because they failed to consider the creation of the universe by God. Thus the earliest naturalists, oblivious of efficient causality, admitted nothing but uncreated matter as the cause of all things; in consequence they had to maintain that matter has always existed. For, since nothing raises itself from non-existence to existence, anything that begins to exist must be caused by something else. And these philosophers asserted that the world has always existed continually, because they recognized only natural agents which were confined to one mode of action, so that the same effect had to follow invariably. Or else they held that the world had an interrupted existence; thus Democritus taught that the world, or rather worlds, had been formed and destroyed over and over again by chance, as a result of the haphazard movements of atoms.

However, it was judged impossible that all the conveniences and utilities existing in nature should be the product of chance, since they are found in all or at least in the majority of cases; still, such a conclusion would necessarily follow if nothing but matter were acknowledged, especially since some effects are observed which are not sufficiently explained by appealing to the causality of matter. Therefore other philosophers postulated an efficient cause, which, according to Anaxagoras, was intelligence[12] or, according to Empedocles, was love and strife.[13] Nevertheless they did not regard such agents as efficient causes of the universe, but rather likened them to other particular agents that act on matter by transforming it from one thing into another. Hence they were forced to hold that matter is eternal, as lacking a cause of its existence. Yet they taught that the world had a beginning, on the score that every effect of a cause which acts by motion follows its cause in duration, since such an effect does not exist until the motion has come to an end; and this end is preceded by the beginning of the movement, with which the agent initiating the movement has to be simultaneous.

On the other hand, Aristotle taught that the world has always existed; for he reflected that, if the efficient cause of the world were held to act by motion, an indefinite recession would ensue, since every movement must be preceded by another movement.[14] His argument did not proceed from the consideration which regards the universe as taking its origin from God, but from the consideration that an agent which begins to act must itself be moved; in other words, he was thinking of a particular cause, not the universal cause. And that is why he derives his

12. Cf. Aristotle, *Phys.*, VIII, 1 (250b 24).
13. Cf. Aristotle, *Phys.*, VIII, 1 (250b 26).
14. This is an Aristotelian argument, as reported by Maimonides, *Guide for the Perplexed*, II, 14 (p. 174).

arguments to prove the eternity of the world from the motion and immobility of the prime mover. If we look closely into the reasons he alleges, they give the impression of being arguments of one who is contending against a definite position. Thus in the beginning of Book VIII of the *Physics*, after introducing the question of the eternity of motion, he starts by reporting the views of Anaxagoras and Empedocles,[15] against which he intends to direct his discussion.

But those who came after Aristotle, judging that the entire universe was produced by God through an act of His will and not by way of motion, endeavored to prove the eternity of the world by arguing that the will does not put off doing what it intends to do unless some innovation or change has intervened. Such an event, at least so we have to imagine, occurs within the flow of time in the one who wills to do something at a definite moment and not earlier.

Nevertheless these men, too, fell into an error similar to that of those who were mentioned above. They thought that the first agent was like any other agent who exercises his activity in time, yet acts through his will. Such an agent is not the cause of time, but presupposes time. God, however, is the cause even of time itself. For time itself is included within the universality of things that have been made by God; therefore, when we speak of the production of all being by God, we ought not to consider the question why He made it at a particular moment rather than sooner. Such a consideration supposes that time preceded the making, instead of being conditional on the making.

But if we consider the production of all creatures in their universality, among which time is included, we ought to inquire why God pre-appointed such and such a measure to time, not why He made the world at such and such a time. The appointing of a measure to time depends on the mere will of God, who decreed that the world should not exist forever but should have a temporal beginning, just as He willed that heaven should not be greater or smaller than it is.

Reply to Objection 1. It is proper to goodness to bring things into existence through an act of the will, whose object is the good. Therefore things did not have to be brought into being in such a way as to be coeternal with the divine goodness; rather they were produced in the way that the divine will disposed for them.

Reply to Obj. 2. Since a heavenly body is incorruptible, it is capable of existing forever. But no power either of existing or of acting regards the past, but only the present or the future; for no one has power over what he has done in the past, since he cannot now cause to have been done what was then left undone. Yet a person has the power to do

---

15. *Phys.*, VIII, 1 (250b 24-29).

such a thing now or in the future. Hence the power of existing forever
that is possessed by heaven does not regard the past, but the future.

Reply to Obj. 3. We cannot say simply that heaven is corruptible
because of the fact that it would lapse into non-existence if it were not
sustained by God. However, we may grant that it is corruptible in a certain sense, that is, in the supposition that God would not sustain it. The
reason for this is that preservation of a creature by God depends on the
divine immutability, not on natural necessity. Consequently no creature
can be said to be absolutely necessary, for it is necessary only on the supposition that the divine will has unchangeably decreed its preservation.

Reply to Obj. 4. Every effect has a necessary relationship to its
efficient cause, whether that cause is natural or voluntary. But our
position is that God is the cause of the world, not by any necessity of His
nature, but by His will, as was explained above.[16] Hence the effect caused
by God must necessarily follow, not so as to be co-extensive in duration
with the divine nature, but at the time disposed for its existence by the
divine will, and precisely such as God willed it to be.

Reply to Obj. 5. The existence of a thing prior to time may be
understood in two ways; first, prior to the whole of time and to everything
that pertains to time. In this sense the world did not exist before time,
because the instant at which the world began, though it is not time, is nevertheless something pertaining to time, not indeed as a part, but as the
starting-point of time. In a second sense a thing is regarded as preceding
time because it precedes complete time; and time is not complete until
the instant which is preceded by another instant. In this sense the world is
prior to time. But that does not necessarily mean that the world is eternal, because even the instant of time which thus precedes time is not eternal.

Reply to Obj. 6. Since every agent produces its like, an effect
must issue from an effectively operating cause in such a way as to retain a
likeness to its cause. But just as an effect produced by a cause that acts by
its nature bears a likeness to the cause, in the sense that it has a form
similar to the form of the agent, so an effect proceeding from a voluntary
agent bears a likeness to the latter, in the sense that it has a form similar
to that of its cause, inasmuch as something is produced in the effect that is
in accordance with the decision of the agent's will. An example of this is
the work produced by a craftsman. Now the will determines not only the
form of the effect, but also its place, duration, and all its conditions.
Therefore the effect produced by the will must come into existence when
the will so determines, not whenever the will itself exists. For the effect is
assimilated to the will, not in being, but in the way the will disposes.

---

16. Cf. art. 15.

Consequently, even though the will should remain the same forever, the effect need not issue from it forever.

Reply to Obj. 7. God necessarily wills His own goodness and everything that is necessarily connected with it. But the production of creatures is not such a thing. Hence the argument is not conclusive.

Reply to Obj. 8. Since God made creatures in order to manifest Himself, it was better and more fitting that they should be produced in such a way as to manifest Him in a more becoming and impressive manner. But God is more impressively manifested by creatures if they have not existed forever; this fact shows clearly that they were brought into existence by another, that God has no need of creatures, and that creatures are completely subject to the divine will.

Reply to Obj. 9. God's will to create the world was, indeed, eternal. He willed, however, that the world should be brought into being, not from eternity, but at the time when He made it. (@time when time began)

Reply to Obj. 10. Before the world existed, it was possible for the world to be made, not indeed by reason of any passive potentiality, but solely by the active power of the agent. Or we may say that it was possible, not because of any potentiality, but because the terms of a proposition such as this: "The world exists," are not mutually contradictory. In this sense a thing is said to be possible without reference to any potentiality as the Philosopher shows in Book V of his *Metaphysics*.[17]

This reply also furnishes the solution to *Objection 11.*

Reply to Obj. 12. The argument proposed envisions an agent that begins to act with a new action; but God's action is eternal, since it is His substance. He is said to begin to act, however, by reason of a new effect which results from His eternal action according to the disposition of His will, which is regarded as the principle of His action with reference to the effect. For an effect results from an action in accord with the condition of any form that is the principle of the action; thus a thing is heated by the heating action of the fire in accord with the degree of heat in the fire.

Reply to Obj. 13. The argument is concerned with an agent that produces an effect in time, yet is not the cause of time. But it does not apply to God, as is clear from what was stated above.[18]

Reply to Obj. 14. If movement is taken literally, the divine will is not moved. Metaphorically, however, it is said to be moved by its object; in this sense, God's goodness alone moves His will, as Augustine points out when he says that God moves Himself independently of place and time.[19] Yet it does not follow that the production of creatures is co-eternal with His goodness, because creatures proceed from God, not as

17. Aristotle, *Metaph.* V, 12 (1019b 34).
18. Art. 15.
19. *De Genesi ad litteram,* VIII, 22 (PL 34, 389).

though they were due or necessary to His goodness, since the divine goodness has no need of creatures and does not stand to gain anything from them, but by His mere will.

Reply to Obj. 15. Since the first succession of time is caused by the succession of movement, as is stated in Book IV of the *Physics,*[20] it is true that every instant is both a beginning and an end of time, as it is also true that every moment is a beginning and an end of motion. Accordingly, if we suppose that movement has not always existed or that it will not always exist, we shall not have to grant that every instant is a beginning and an end of time; for some instant will be only the beginning and some will be only the end. Clearly, therefore this argument goes around in a circle, and consequently it is not a demonstration; yet it is effective for the purpose intended by Aristotle, who advances it against a definite position, as was mentioned in the body of this article. Many arguments are thus effective for attacking a position because of statements made by adversaries, and yet they are not absolutely sound.

Reply to Obj. 16. An instant is, indeed, always regarded as flowing; not, however, as always flowing from one point to another, but sometimes as flowing only *from* a point, such as the last instant of time, and sometimes as flowing only *toward* a point, as the first instant.

Reply to Obj. 17. The argument in question does not prove that motion has endured forever, but that circular motion can endure forever, because no certain conclusion about motion can be derived from mathematics. Hence Aristotle does not prove the eternity of motion by arguing that it is circular; but, on the supposition that motion is eternal, he shows that it is circular, because no other motion can be eternal.[21]

This also provides an answer to *Objection 18.*

Reply to Obj. 19. The knowable is related to our knowledge as God's knowledge is related to creatures. For God's knowledge is the cause of creatures, as the knowable is the cause of our knowledge. Consequently, just as the object of knowledge can exist even if we have no knowledge of it, as is stated in the *Categories,*[22] so God can have knowledge even if the knowable thing does not exist.

Reply to Obj. 20. God precedes the world in duration, not indeed of time, but of eternity, because God's existence is not measured by time. Yet, prior to the world there was no real time, but only imaginary time; that is, we are able at the present moment to imagine that infinite eons of time, co-existent with eternity, could have been coursing along before the beginning of time.

---

20. Aristotle, *Phys.,* IV, 11, (219a 19-25).
21. *Phys.,* VIII, 9 (265a 25).
22. Aristotle, *Cat.,* 7 (7b 30).

Reply to Obj. 21. If the relation of lordship is regarded as consequent on the action whereby God actually governs creatures, then God was not Lord from eternity. But if it is regarded as consequent on His power of governing, it belongs to Him from eternity. So we do not have to conclude that creatures existed from eternity, except potentially.

Reply to Obj. 22. Augustine employs this argument to prove the Son's co-eternity and co-equality with the Father.[23] But the argument is not applicable to the world, because the Son's nature is identical with the Father's, and therefore has to be co-eternal and co-equal with that of the Father; withholding of such perfections would savor of envy. A creature's nature, however, does not require these perfections; hence the comparison fails.

Reply to Obj. 23. According to the Greek version, the text reads: "The Lord shall reign for age upon age, and beyond." In a commentary on this passage, Origen explains that "age" means a period of one generation whose end is known to us, while "age upon age" refers to an immense span of time which has, indeed, an end, though it is unknown to us; and God's reign will extend even "beyond" that.[24] Accordingly, "eternity" here signifies a long time. Anselm, however, in his *Proslogium,*[25] takes eternity to mean aeviternity, which never comes to an end; and yet God is said to reign beyond it, for three reasons: first, because aeviternal beings can be thought of as not existing; secondly, because they would not exist if God did not sustain them in being, so that of themselves they have no existence; thirdly, because they do not have their whole existence at once, but are subject to some successive change.

Reply to Obj. 24. A thing which begins to exist must have a measure of its duration to the extent that it begins through motion. But time does not begin in this way by creation, and so the argument proves nothing. Besides, every measure may be said to be measured by itself within its own genus, as a line is measured by a line, and likewise time is measured by time.

Reply to Obj. 25. Time is not like permanent things which possess their substance all at once; hence the whole of time does not have to exist as soon as it begins. Consequently there is nothing to prevent us from saying that time begins in an instant.

Reply to Obj. 26. God's action is eternal, but the effect caused by it is not eternal, as was shown above. Therefore, even though God had not always been a cause, inasmuch as there was not always an effect, it does not follow that He was not a cause potentially, since His action has

23. *Contra Maximinum,* II, cc. 7; 23 (PL 42, 761 f; 796-801).
24. Origen, *In Exodum,* hom. VI, 13 (PG 12, 340).
25. Cap. 20 (PL 158, 237; ed. Schmitt, I, 115).

always existed; however, the conclusion follows if the potentiality in question refers to the effect.

Reply to Obj. 27. According to the Philosopher, truth is in the mind, not in things,[26] for truth is the conformity of the intellect to things. Hence all that has been true from eternity has been true by the truth present in the divine intellect, which is eternal.

Reply to Obj. 28. All things which are said to be true from eternity are true, not by different kinds of truth, but by one and the same truth existing in the divine intellect, with relationship, however, to various things as future in their own proper existence. And so some distinction in that truth can be assigned in accordance with the diversity of the relationship.

Reply to Obj. 29. The Philosopher's assertion refers to our mental or oral statements; for the truth found in our mind or in our utterance is caused by existing things. Conversely, however, truth in the divine intellect is the cause of things.

Reply to Obj. 30. On the part of God Himself, *to make* and *to speak* are not in any way different, for God's action is not an accident, but is His substance; nevertheless *to make* implies an effect actually existing in its own proper nature, whereas this is not implied by *to speak.* The arguments proposed in favor of the contrary position do indeed conclude, though not with necessity, except the first, which is based on authority. The argument about intellectual keenness that increases in the course of time does not prove that time had a beginning. For study of the sciences could have been interrupted repeatedly, and later, after long intervals, have been taken up afresh, as the Philosopher also remarks.[27] Furthermore, although the earth may be washed away by floods in one part, it is augmented in other parts by the mutual conversion of the elements. Finally, although God's duration is identical with his nature in reality, it differs from His nature logically. Therefore the fact that He is prior to the world by nature does not necessarily lead to the conclusion that He is prior to it in duration.

---

26. Aristotle, *Metaph.,* VI, 4 (1027b 25).
27. Aristotle, *Metaph.,* XI, 12 (1068a 31).

St. Thomas Aquinas

Summa Theologiae

Part I, Question XLVI, Articles 1 and 2

Article 1

Whether the Created Universe Has Existed Forever

We proceed to the First Article as follows. It may seem that the created universe, which is now called the world, had no beginning, but existed from eternity.

1. For everything that begins to exist had the possibility of existing before it existed; otherwise it could not have come into existence. If, therefore, the world began to exist, it had the possibility of existing before it began. But that which has the possibility of existing is matter; for matter is in potentiality to existence, which is brought about by a form, and to non-existence, which is the result of privation. Therefore, if the world began to exist, matter must have existed before the world. But matter cannot exist without form; and the world's matter united to form is the world. Therefore the world existed before it began to exist; which is impossible.

2. Besides, nothing that has the power to exist forever, at some time exists and at some time does not exist, because a thing exists as long as its power to exist endures. But every incorruptible thing has the power to exist forever, for its power is not limited to any particular span of duration. Therefore no incorruptible thing at some time does not exist. But everything that begins to exist, at some time exists and at some time does not exist. Therefore no incorruptible thing begins to exist. But there are many incorruptible things in the world, such as the heavenly bodies and all intellectual substances. Therefore the world did not begin to exist.

3. Moreover, what is ungenerated did not begin to exist. But the Philosopher says, in *Physics* I, that matter is ungenerated;[1] and in *De caelo* I, he asserts that heaven is ungenerated.[2] Therefore the universe had no beginning.

---

1. Aristotle, *Phys.,* I, 9 (192a 28).
2. *De caelo,* 1, 3 (270a 13).

4. Further, there is no body in a vacuum, although there could be. But if the world had a beginning, there was formerly no body where the body of the world is now; and yet a body could have been there, as otherwise it would not be there now. Therefore prior to the world there was a vacuum; which is impossible.

5. Besides, nothing begins to be moved anew, unless a mover or a movable thing is in a different condition now than it was before. But what is in a different condition now than it was before, is moved. Therefore, prior to all new motion, there was some previous motion. Accordingly there was always some motion. And therefore there was always some movable thing, because motion is not found except in a movable thing.

6. Moreover, every mover is either natural or voluntary. But neither begins to move except in response to some pre-existing motion. For nature always works in the same way; consequently, unless some change either in the nature of the mover or in the movable thing precedes, no motion that was not in operation before will be inaugurated by a natural mover. And as for the will, it may postpone doing what it proposes to do without undergoing any change; but even in this case, some sort of change is imagined, at least as regards the passage of time. Thus a person who wishes to build a house tomorrow and not today, waits for something to happen tomorrow that does not occur today, at the very least he awaits the passing of the present day and the coming of the morrow; and this involves change, because time is the measure of motion. Therefore, prior to all newly beginning motion, there was some other motion. And so we have the same conclusion as before.

7. Further, whatever is always in its beginning and always in its end, can neither cease nor begin; for what begins is not in its end, and what ceases is not in its beginning. But time is always in its beginning and its end, since no time exists except *now* , which is the end of the past and the beginning of the future. Therefore time can neither begin nor end. Consequently motion, too, cannot begin or end, for time is its measure.

8. Besides, God is prior to the world either in the order of nature alone, or also in the order of duration. If in the order of nature alone, then, since God exists from eternity, the world also exists from eternity. And if He is prior in duration, then, since before and after in duration constitute time, time existed before the world; and that is impossible.

9. Further, given a sufficient cause, its effect follows; for a cause that is not followed by an effect is an imperfect cause, needing something else to produce the effect. But God is the sufficient cause of the world. He is the final cause, by reason of His goodness; the exemplary cause, by reason of His wisdom; and the efficient cause, by reason of His power, as

is clear from the preceding exposition.[3] Since, therefore, God is eternal, the world has also existed from eternity.

10. Moreover, if an action is eternal, its effect is likewise eternal. But God's action is His substance, which is eternal. Therefore the world, too, is eternal.

ON THE CONTRARY, however, we read in John 17:5: "Glorify Thou Me, O Father, with Thyself, with the glory which I had before the world was"; and in Proverbs 8:22: "The Lord possessed Me in the beginning of His ways, before He made anything from the beginning."

I ANSWER THAT nothing has existed from eternity except God. And this position is far from impossible. It was shown above that God's will is the cause of things.[4] Therefore some things necessarily exist according as God necessarily wills them, as is stated in *Metaphysics* V.[5] But it was shown above that, absolutely speaking, God is under no necessity to will anything except Himself.[6] Therefore it is not necessary for God to will that the world should have existed forever. All we can say is that the world is eternal if God wills it to be so, since the existence of the world depends on God's will as on its cause. Consequently it is not necessary for the world to exist always. Hence no demonstrative proof is possible.

As for the arguments which Aristotle brings forward in support of this contention, they are not absolutely demonstrative, but only relatively so. That is, they serve to overthrow the arguments of ancient philosophers who held that the world began in certain ways that are in truth impossible. This appears from three considerations. First, because both in *Physics* VIII[7] and in *De caelo* I[8] he premises certain opinions, such as those of Anaxagoras, Empedocles, and Plato, and proposes arguments to refute them. Secondly, whenever he treats of this matter, he reports the testimony of the ancients, which is not the procedure of one who undertakes a demonstration, but rather of one who uses probable arguments to persuade. Thirdly, he expressly says in Book I of the *Topics* that there are some dialectical problems which we have no means of solving; one of these is, whether the world is eternal.[9]

Reply to Objection 1. Before the world existed, it was possible for the world to exist, not indeed by reason of passive potentiality, which is matter, but by reason of the active power of God. And the world was possible also in the sense that a thing is said to be absolutely possible, not

---

3. Q. 44, aa. 1, 3, 4.
4. Q. 19, a.4.
5. Aristotle, *Metaph.,* V, 5 (1015b 9).
6. Q.19, a.3.
7. *Phys.,* VIII, 1 (250b 24; 251b 17).
8. *De caelo,* I, 10 (279b 4; 280a 30).
9. *Top.,* I, 9 (104b 16).

by virtue of some potentiality, but from the sole relation of the terms which are not mutually incompatible. In this sense possible is opposed to impossible, as is stated by the Philosopher in *Metaphysics* V.[10]

Reply to Obj. 2. That which has the power to exist always, does not indeed, once it has that power, sometimes exist and sometimes not exist. However, before it had that power, it did not exist. Hence this argument, which is proposed by Aristotle in *De caelo* I,[11] does not prove absolutely that incorruptible things never began to exist, but only that they did not begin to exist according to the natural process by which things subject to generation and corruption begin to exist.

Reply to Obj. 3. Aristotle proves in *Physics* I that matter is ungenerated from the fact that it has no subject from which to derive its existence.[12] And in *De caelo* I he proves that the heavens are ungenerated because they have no contrary from which to be generated.[13] Clearly, then, no conclusion follows from either argument except that matter and the heavens did not begin by way of generation, as some have asserted, especially with regard to the heavens.[14] But our contention is that matter and the heavens were brought into being by creation, as was shown above.[15]

Reply to Obj. 4. The notion of a vacuum implies more than that in which nothing is present; it requires space which is capable of containing a body but does not contain one, as is clear from Aristotle in *Physics* IV.[16] However, we hold that there was no place or space prior to the world.

Reply to Obj. 5. The first mover was always in the same state, but the first movable thing was not always in the same state, because it began to exist, whereas previously it had not existed. However, this was brought about, not by change, but by creation, which is not change, as was shown above.[17] Hence it is evident that this argument, which Aristotle gives in *Physics* VIII,[18] scores a point against those who acknowledged eternal movable things but not eternal motion, as maintained in the opinions of Anaxagoras and Empedocles.[19] Our position, however, is that there was always motion from the instant that movable things began to exist.

10. Aristotle, *Metaph.*, V, 12 (1019b 19).
11. *De caelo*, I, 12 (281b 18).
12. *Phys.*, I, 9 (192a 28).
13. *De caelo*, I, 3 (270a 13).
14. *Ibid.*, I, 10 (279b 13).
15. Q.45, a.2.
16. *Phys.*, IV, 1 (208b 26).
17. Q.45, a.2, ad2.
18. *Phys.*, VIII, 1 (251a 25).
19. Cf. *ibid.* (250b 24).

Reply to Obj. 6. The first agent is a voluntary agent. And, although He had an eternal will to produce some effect, He did not produce an eternal effect. Nor does some change have to be presupposed, not even with regard to imaginary time. We must attend to the differences between a particular agent, that presupposes one thing and causes another, and the universal agent, that produces the whole effect. A particular agent produces a form, and presupposes matter; such an agent must introduce a form that is proportionate to suitable matter. With good reason, therefore, we assert that a particular agent introduces a form into some portion of matter and not into other matter, because matter differs from matter. But we are not reasonable if we assert the same of God, who produces form and matter together; rather, we are reasonable when we say that God produces matter suitable for the form and the end. Now a particular agent presupposes time no less than matter. Such an agent is reasonably described as acting in later time and not in earlier time, according as we imagine the passage of time succeeding time. But the universal agent, who produces both the thing and time, is not to be regarded as acting now and not before, according to an imaginary passage of time succeeding time, as though time were presupposed to His action; the truth is, that He assigned to His effect as much time as He willed, as long a period as was suitable for showing forth his power. For if the world has not always existed, it leads us more clearly to a knowledge of the divine creative power than if it had always existed. The reason is, that everything that has not always existed evidently has a cause; but this is not so evident in the case of a thing that has existed forever.

Reply to Obj. 7. As is stated in *Physics* IV, "before" and "after" pertain to time, seeing that "before" and "after" are found in motion. Hence beginning and end in time are to be understood in the same way as in motion.[20] In the hypothesis of the eternity of motion, any given moment in motion must be a beginning and an end of motion; but this is not necessary if motion has a beginning. And the same is true of the now of time. And thus it is clear that the idea of the instant now, viewed as being always the beginning and end of time, presupposes the eternity of time and motion. Hence Aristotle brings in this argument in *Physics* VIII against those who affirmed the eternity of time, but denied the eternity of motion.[21]

Reply to Obj. 8. God is prior to the world in duration. Here, however, the word "prior" signifies priority, not of time, but of eternity. Or we may say that it signifies the eternity of imaginary time, not of real time. In the same way, when we say that there is nothing above the heavens, the word "above" designates only an imaginary place, according as we

---

20. *Ibid.*, IV, 11 (219a 17).
21. *Ibid.*, VIII, 1 (251b 29.).

are able to imagine other dimensions over and above the dimensions of the heavenly body.

Reply to Obj. 9. As an effect follows from a cause that acts by nature in accordance with the kind of form it has, so likewise an effect follows from a voluntary agent in accordance with the form preconceived and determined by him, as is clear from what we said above.[22] Therefore, although God was from eternity the sufficient cause of the world, we do not have to hold that the world was produced by Him otherwise than as preordained by His will, that is, that it should have existence after non-existence, so as more clearly to manifest its Author.

Reply to Obj. 10. Once an action is performed, its effect follows as required by the form which is the principle of the action. In the case of voluntary agents, that which is conceived and preordained is regarded as the principle of the action. From God's eternal action, consequently, there follows, not an eternal effect, but an effect such as God willed it to be, that is, one which has existence subsequent to non-existence.

Article 2

Whether it is an Article of Faith that the World Had a Beginning

We proceed to the Second article as follows. It may seem that it is not an article of faith but rather a demonstrable conclusion that the world began.

1. For everything that is made has a beginning of its duration. But it can be proved demonstratively that God is the efficient cause of the world; and this was also the conviction of some of the more approved philosophers. Therefore it can be proved demonstratively that the world had a beginning.

2. Besides, if we have to say that the world was made by God, then it must have been made either from nothing or from something. But it was not made from something, because in that case the matter of the world would have preceded the world itself; opposed to this position are the arguments advanced by Aristotle, who held that the heavens are ungenerated. Therefore we must conclude that the world was made from nothing. And thus it has existence coming after nonexistence. Therefore it must have had a beginning.

---

22. Q.19, a.4; q.41, a.2.

3. Further, everything that acts intellectually, begins its action from some starting point,[23] as we observe in all works of art. But God acts by His intellect. Therefore He begins His action from some starting point. Consequently the world, which is His effect, did not always exist.

4. Moreover, we perceive clearly that certain arts and the inhabiting of certain regions of the earth began at definite times. But this would not be true if the world has existed forever. Evidently, therefore, the world has not existed forever.

5. Further, it is certain that nothing can be equal to God. But if the world had always existed, it would be equal to God in duration. Therefore it is certain that the world has not always existed.

6. Besides, if the world has always existed, an infinite number of days has preceded the present day. But infinity cannot be traversed. Therefore we should never have arrived at the present day - which is obviously false.

7. Further, if the world has existed eternally, generation has also taken place eternally. Therefore one man has been begotten by another man in an infinite series. But the father is the efficient cause of his son, as is stated in *Physics* II.[24] Therefore efficient causes are linked in an infinite series - which is rejected in *Metaphysics* II.[25]

8. Moreover, if the world has always existed and generation has always taken place, an infinite multitude of men would have preceded us. But man's soul is immortal. Therefore an infinite multitude of human souls would now be actually existing - which is impossible. Therefore we can know with strict certainty that the world had a beginning; and so this truth is not held by faith alone.

ON THE CONTRARY, however, articles of faith cannot be proved demonstratively, because faith is concerned with "things that appear not," according to Hebrews 11:1. But that God is the Creator of the world in such a way that the world began to exist, is an article of faith; for we say: "I believe in one God," etc.[26] And likewise Gregory points out that Moses prophesied about the past when he said, "In the beginning God created heaven and earth."[27] In these words the inception of the world is conveyed. Therefore the inception of the world is known exclusively by revelation. Accordingly it cannot be proved demonstratively.

I ANSWER that we hold by faith alone that the world has not existed forever; this truth cannot be proved demonstratively. We spoke in

---

23. Cf. Aristotle, *Phys.*, III, 4 (203a 31).
24. *Ibid.*, II, 3 (194b 30).
25. Aristotle, *Metaph.*, II, 2 (994a 5).
26. Nicene Creed (Denz., 54).
27. Gregory the Great, *In Ezechielem*, I, 1 (PL, 76, 786).

the same way above, in connection with the mystery of the Trinity.[28]
The reason is that the inception of the world cannot be demonstrated by
arguing from the standpoint of the world itself. For the principle of
demonstration is the essence of a thing. But everything, viewed in its
species, abstracts from *here* and *now* ; this is why universals are said to be
everywhere and always.[29] Hence it cannot be demonstrated that man or
the heavens or a stone did not always exist.

Likewise, this truth cannot be demonstrated by arguing from the
standpoint of the efficient cause, which acts voluntarily. For God's will
cannot be investigated by reason, except as regards those matters which
God must will with absolute necessity; such however, are not those things
which He wills with reference to creatures, as we said above.[30] But the
divine will can be made known to man by revelation, on which faith is
based. That the world had a beginning, therefore, is an object of faith,
but not of demonstration or science. And we do well to keep this in mind;
otherwise, if we presumptuously undertake to demonstrate what is of
faith, we may introduce arguments that are not strictly conclusive; and
this would furnish infidels with an occasion for scoffing, as they would
think that we assent to truths of faith on such grounds.

Reply to Objection 1. As Augustine says, philosophers who
affirmed the eternity of the world were divided between two opinions.[31]
Some asserted that the substance of the world does not come from God.
And this is an intolerable error, which is refuted by cogent proofs.
Others, however, who advocated the eternity of the world, nevertheless
granted that it was made by God. "For they contend that the world had a
beginning, not of time, but of its creation, so that, in some scarcely intelligible sense, it was made, yet from eternity." And St. Augustine continues:
"They found a way of explaining their view. If, they say, a foot had always,
from eternity, been planted in the dust, there would always be a footprint
underneath, and no one would doubt that it has been made by someone
stepping there. In the same way the world has always existed, since He
who made it has always existed."[32] To understand this, we must reflect
that an efficient cause which acts by motion necessarily precedes its effect
in the order of time, because the effect does not exist until the end of the
action, and every such agent must be an initiator of action. But if the
action is instantaneous and not successive, the cause does not have to precede the effect in duration, as is exemplified in the case of illumination.

---

28. Q.32, a.1.
29. Aristotle, *Posterior Analytics,* I, 31 (87b 33).
30. Q.19, a.3.
31. *De civitate Dei,* XI, 4 (PL, 41, 319).
32. *Ibid.,* X, 31 (PL, 41, 311).

Hence they insist that, even though God is the active cause of the world, it does not necessarily follow that He precedes the world in duration,[33] because creation by which He produced the world, is not a successive change, as was said above.[34]

Reply to Obj. 2. Advocates of the theory that the world is eternal would say that the world was made by God from nothing, not in the sense that it was made after nothing, according to what we understand by the word, creation, but because it was not made from anything. And so some of them do not repudiate the term, creation, as we learn from Avicenna in his *Metaphysics.*[35]

Reply to Obj. 3. This is the argument of Anaxagoras, as given in *Physics* III.[36] But it does not conclude with necessity, except in the case of an intellect that deliberates in order to find out what ought to be done; and this procedure is similar to motion. Such is the human intellect, but not the divine intellect, as was shown above.[37]

Reply to Obj. 4. Those who hold the eternity of the world suppose that some region was uninhabitable, then inhabitable, and so on, back and forth, in an infinite succession of changes.[38] They likewise postulate that the arts, because of various catastrophies and accidents, were discovered and later on decayed again in infinite succession.[39] Aristotle observes, in his *Meterology,* that it is ridiculous to admit the theory about the inception of the entire world on the basis of particular changes such as these.[40]

Reply to Obj. 5. Even if the world had always existed, it would not be equal to God in eternity, as Boethius says at the end of his *Consolation.*[41] For the divine existence is existence that is simultaneously whole, without succession; but such is not the case with the world.

Reply to Obj. 6. Passage is always understood as proceeding from one term to another. No matter what past day may be designated, a finite number of days has elapsed between it and the present day, and these days can be traversed. The objection is based on the supposition that infinite intervals lie between given extremes.[42]

---

33. Cf. Averroes, *Tahafut al-Tahafut,* I, 65 (*The Incoherence of the Incoherence,* trans. by S. Van den Bergh, [London: Luzac, 1954], I, p. 37).

34. Q.45, a.2, ad 3.

35. *Metaph.,* IX, 4 (ed. Venetiis, 1508, 104 va).

36. Aristotle, *Phys.,* III, 4 (203a 31).

37. Q.14, a.7.

38. Cf. St. Augustine, *De civitate Dei,* XII, 10 (PL, 41, 358).

39. *Ibid.*

40. *Meteor.,* I, 14 (352a 26; 351b 8).

41. *De consolatione philosophiae,* V, pros. 6 (PL, 63, 859).

42. Not all are satisfied with this reply. L. Roy, "Note philosophique sur l'idée de commencement dans la création," *Sciences Ecclésiastiques,* II (1949), 224, complains that St. Thomas fixes a point of departure, and precisely this is in question; to solve the problem, one must not fix an initial day.

Reply to Obj. 7. In efficient causes, a series that is *per se* infinite is impossible, that is, if the causes that are *per se* required for a certain effect should be infinite in number; for example, if a stone is moved by a stick, the stick by a hand, and so on endlessly. But a series of efficient causes that is *per accidens* infinite is not considered impossible, that is, if all the causes that are infinitely multiplied are in the order of a single cause, while their multiplication is accidental, as when a craftsman works with many hammers *per accidens,* because one after another breaks. Thus it is accidental that any particular hammer should act after the action of another hammer; and likewise it is accidental that any particular man, in his capacity as begetter, should have been begotten by another man. He begets as a man, not as the son of another man; for all men who beget function as units in the order of efficient causes; each functions as a particular begetter. Hence it is not impossible that man should be begotten by man in an infinite series. But such a state of affairs would be impossible if the generative action of this man depended on that man and on an elementary body and on the sun, and so on to infinity.

Reply to Obj. 8. Those who assert the eternity of the world evade this argument in various ways. Some of them think that the actual existence of an infinite multitude of souls is not an impossibility, as we learn from the *Metaphysics* of Algazel, who says that such infinity is accidental.[43] But this contention was refuted above.[44] Some say that the soul is corrupted along with the body.[45] And some affirm that of all souls only one remains.[46] Yet others, as Augustine reports, solved the difficulty by proposing a rotation of souls; that is, after definite periods of time, souls that have been separated from their bodies again return to bodies.[47] These opinions will all be treated later.[48] For the present, we may note that this argument is limited to a particular class of beings. Hence one might say that the world, or at any rate some creature, such as an angel, is eternal, even though man is not eternal. Our intention, however, was to discuss the general question, whether any creature has existed from eternity.

---

43. *Metaph.,* I, tr. 1, div. 6, (ed. J. T. Muckle, [Toronto: St. Michael's College, 1933], p. 40 f.).
44. Q.7, a.4.
45. Thus Democritus, Epicurus, and others; cf. Theodoret of Cyrus, *Graecarum affectionum curatio,* serm. V. (PG, 83, 932).
46. Averroes, *Tahafut al-Tahafut,* I, 29 (Van den Bergh, p. 15).
47. Serm. 241, 4 (PL, 38, 1135); *De civit. Dei,* XII, 13 (PL, 41, 361).
48. Q.75, a.6; q.76, a.2; q.118, a.3.

St. Thomas Aquinas

Compendium of Theology
(Compendium Theologiae)

Part One, Chapters 98 and 99

Chapter 98

Question of the Eternity of Motion

We might imagine that, although God can produce a new effect by His eternal and immutable will, some sort of motion would have to precede the newly produced effect. For we observe that the will does not delay doing what it wishes to do, unless because of some motive that is operative now but will cease later, or because of some motive that is inoperative now but is expected to become operative in the future. In summer a man has the will to clothe himself with a warm garment, which, however, he does not wish to put on at present, but in the future; for now the weather is warm, although it will cease to be warm with the advent of a cold wave later in the year. Accordingly, if God wished from eternity to produce some effect, but did not produce it from eternity, it seems either that something was expected to happen in the future that had not yet occurred, or else that some obstacle had to be removed that was then present. Neither of these alternatives can take place without motion. Thus it seems that a subsequent effect cannot be produced by a preceding will unless some motion previously occurs. And so, if God's will relative to the production of things was eternal, and nevertheless things were not produced from eternity, their production must have been preceded by motion, and consequently by mobile objects. And if the latter were produced by God, but not from eternity, yet other motions and mobile objects must have preceded, and so on, in infinite recession.

The solution to this objection readily comes to mind if we but attend to the difference between a universal and a particular agent. A particular agent has an activity that conforms to a norm and measure prescribed by the universal agent. This is clear even in civil government. The legislator enacts a law which is to serve as a norm and measure. Any particular judge must base his decisions on this law. Again, time is the measure of actions which occur in time, so that he acts for some definite reason now, and not before. But the universal agent, God, instituted this measure, which is time, and He did so in accord with His will. Hence time also is to be numbered among the things produced by God. Therefore, just as the quantity and measure of each object are such as God wishes to assign to it, so the quantity of time is such as God wished to mete out; that is, time and the things existing in time began when God wished them to begin.

The objection we are dealing with argues from the standpoint of an agent that presupposes time and acts in time, but did not institute time. Hence the question, why God's eternal will produces an effect now and not earlier, presupposes that time exists; for "now" and "earlier" are segments of time. With regard to the universal production of things, among which time is also to be counted, we should not ask: "Why now and not earlier?" Rather we should ask: "Why did God wish this much time to intervene?" And this depends on the divine will, which is perfectly free to assign this or any other quantity to time. The same may be noted with respect to the dimensional quantity of the world. No one asks why God located the material world in such and such a place rather than higher up or lower down or in some other position; for there is no place outside the world. The fact that God portioned out so much quantity to the world that no part of it would be beyond the place occupied in some other locality, depends on the divine will. However, although there was no time prior to the world and no place outside the world, we speak as if there were. Thus we say that before the world existed there was nothing except God, and that there is no body lying outside the world. But in thus speaking of "before" and "outside," we have in mind nothing but time and place as they exist in our imagination.

Chapter 99

Controversy on the Eternity of Matter

However, even though finished products were not in existence from eternity, we might be inclined to think that matter had to exist from eternity. For everything that has being subsequent to non-being, is changed from non-being to being. Therefore if created things, such as heaven and earth and the like, did not exist from eternity, but began to be after they had not been, we must admit that they were changed from non-being to being. But all change and motion have some sort of subject; for motion is the act of a thing existing in potency. However, the subject of the change whereby a thing is brought into existence, is not the thing itself that is produced, because this thing is the terminus of the motion, and the terminus and subject of motion are not the same. Rather, the subject of the change is that from which the thing is produced, and this is called matter. Accordingly, if things are brought into being after a state of non-being, it seems that matter had to exist prior to them. And if this matter is, in turn, produced subsequent to a period of non-existence, it had to come from some other, pre-existing matter. But infinite procession along these lines is impossible. Therefore we must eventually come to eternal matter, which was not produced subsequent to a period of non-existence.

Again, if the world began to exist after it had first not existed, then, before the world actually existed, it was either possible for the world to be or become, or it was not possible. If it was not possible for the world to be or to become, then, by equipollence, it was impossible for the world to be or to become. But if it is impossible for a thing to become, it is necessary for that thing not to become. In that case we must conclude that the world was not made. Since this conclusion is patently false, we are forced to admit that if the world began to be after it had first not been, it was possible for it to be or to become before it actually existed. Accordingly there was something in potency with regard to the becoming and being of the world. But what is thus in potency to the becoming and existence of something, is the matter of that something, as we see exemplified in the case of wood relative to a bench. Apparently, therefore, matter must have existed always, even if the world did not exist always.

As against this line of reasoning, we showed above that the very matter of the world has no existence except from God.[1] Catholic faith does not admit that matter is eternal any more than it admits that the world is eternal. We have no other way of expressing the divine causality in things themselves than by saying that things produced by God began to exist after they had previously not existed. This way of speaking evidently and clearly brings out the truth that they have existence not of themselves, but from the eternal Author.

The arguments just reviewed do not compel us to postulate the eternity of matter, for the production of things in their totality cannot properly be called change. In no change is the subject of the change produced by the change, for the reason rightly alleged by the objector, namely, that the subject of change and the terminus of the change are not identical. Consequently, since the total production of things by God, which is known as creation, extends to all the reality that is found in a thing, production of this kind cannot properly verify the idea of change, even though the things created are brought into existence subsequently to non-existence. Being that succeeds to non-being does not suffice to consti-

God doesn't do any work to create

---

1. Chap. 69.

tute real change, unless we suppose that a subject is first in a state of privation, and later under its proper form. Hence "this" is found coming after "that" in certain things in which motion or change do not really occur, as when we say that day turns into night. Accordingly, even though the world began to exist after having not existed, this is not necessarily the result of some change. In fact, it is the result of creation, which is not a true change, but is rather a certain relation of the created thing, as a being that is dependent on the Creator for its existence and that connotes succession to previous non-existence. In every change there must be something that remains the same although it undergoes alteration in its manner of being, in the sense that at first it is under one extreme and subsequently under another. In creation this does not take place in objective reality, but only in our imagination. That is, we imagine that one and the same thing previously did not exist, and later existed. And so creation can be called change, because it has some resemblance to change.

The second objection, too, lacks cogency. Although we can truly say that before the world was, it was possible for the world to be or to become, this possibility need not be taken to mean potentiality. In propositions, that which signifies a certain modality of truth, or in other words that which is neither necessary nor impossible, is said to be possible. What is possible in this sense does not involve any potentiality, as the Philosopher teaches in Book V of his *Metaphysics*.[2] However, if anyone insists on saying that it was possible for the world to exist according to some potency, we reply that this need not mean a passive potency, but can mean active potency; and so if we say that it was possible for the world to be before it actually was, we should understand this to mean that God could have brought the world into existence before He actually produced it. Hence we are not forced to postulate that matter existed before the world. Thus Catholic faith acknowledges nothing to be coeternal with God, and for this reason professes that He is the "Creator and Maker of all things visible and invisible."

Paper topic
What's the diff between
who is the God who is
who is the God who is the
Creator, & who is the God
biggest thing around.

---

2. *Metaph.*, V, 12 (1019b 35).

BIBLIOGRAPHY

Averroes' Tahafut al-Tahafut. Translated from the Arabic by S. van den Bergh. 2 vols. London: Luzac, 1954. Introduction, vol. I, pp. ix-xxxvi.

de Blic, J., "Les arguments de saint Augustin contre l'éternité du monde," *Mélanges de Science Religieuse, II (1945), 33-34.*

————. "A propos de l'éternité du monde," *Bulletin de Littérature Ecclesiastique,* XLVII (1946), 162-70.

Boehner, P., "On the Production of Creatures," in *St. Thomas Aquinas, Summa Theologica,* Literally translated by Fathers of the English Dominican Province. 3 vols. New York: Benziger, 1948, III, 3174-86.

Chenu, M. D., *Introduction à l'étude de saint Thomas d'Aquin.* Montreal, Paris: Institut d'études médiévales, 1950.

De Haes, P., "Creatio in tempore," *Collectanea Mechliniensia,* XXXVI (1951), 585-90.

Gierens, M., *Controversia de aeternitate mundi,* Textus et documenta, Series philosophica 6 Romae: Pontificia Universitas Gregoriana, 1933.

Gillon, L. B., "Thomas d'Aquin. V. Signification historique de la théologie de saint Thomas," *Dictionnaire de Théologie Catholique,* XV, col. 663-72.

Gilson, E., *The Christian Philosophy of St. Thomas Aquinas.* With a Catalogue of St. Thomas' Works by I. T. Eschmann, O.P. Translated by L. K. Shook, C.S.B. New York: Random House, 1956.

————. *History of Christian Philosophy in the Middle Ages.* New York: Random House, 1955.

Grison, M., *Problèms d'origines: l'univers, les vivants, l'homme.* Paris: Letousey et Ané, 1954.

Landucci, P. C. "Si può dimonstrare filosoficamente la temporaneità a finitezza dell' universo materiale?" *Divus Thomas* (Piac.), LII (1949), 340-44.

————. "L'infinità dimensiva e temporale dell' universo e veramente assurda," *Divus Thomas* (Piac.), LIV (1951), 60-79.

Mondreganes, A., "De impossibilitate aeternae mundi creationis ad mentem S. Bonaventurae," *Collectanea Franciscana,* V (1935), 529-70.

de Munnynck, M., "Le commencement du monde," *Divus Thomas* (Fri), IV (1926), 33-39.

Pelster, F., "Zur Echtheit der Concordantia dictorum Thomae und zur Datierung von De aeternitate mundi," *Gregorianum,* XXXVII (1956), 610-22.

Pinard, H., "Creation," *Dictionnaire de Thèologie Catholique,* III, col. 2034-2201.

Roy, L., "Note philosophique sur l'idée de commencement dans la création," *Sciences Ecclésiastiques* II (1949), 220-25.

Sertillanges, A. D., *L'idée de création et ses retentissements en philosophie.* Paris: Aubier, 1945.

Van Steenberghen, F., *The Philosophical Movement in the Thirteenth Century.* Edinburgh: Nelson, 1955.

Vollert, C., "Creation," in *St. Thomas Aquinas, Summa Theologica,* Literally translated by Fathers of the English Dominican Province, 3 vols. New York: Benziger, 1948. III, 3164-73.

_____. "Origin and Age of the Universe Appraised by Science," *Theological Studies,* XVIII (1957), 137-68.

Wolfson, H. A., *Philo: Foundations of Religious Philosophy in Judaism, Christianity, and Islam.* 2 vols. Cambridge, Mass: Harvard University Press, 1947.

Siger of Brabant

*On the Eternity
of the World*

(De Aeternitate Mundi)

Translated from the Latin
With an Introduction

By

LOTTIE H. KENDZIERSKI, PH.D.
Professor of Philosophy
Marquette University

Translator's Introduction

Siger of Brabant's treatise,[1] the *De Aeternitate Mundi,* has been discovered in four different manuscripts.[2] Manuscript A is the *Paris Nat. Latin* , 16222, discovered by C. Potvin in 1878.[3] Manuscript B is the *Paris Nat. Latin,* 16297, discovered by P. Mandonnet during his preparation of Manuscript A.[4] Mandonnet published Manuscript A and added an Appendix which consisted of a short description of Manuscript B, as well as variations in the manuscripts. In a later edition of the text, Mandonnet revised his method and published Manuscript B with variations from Manuscript A. In the later edition, Mandonnet considered Manuscript B to be superior; Manuscript A became a report of a student.[5] Manuscript C, discovered by F. Pelster in 1925, showed a marked resemblance to Manuscript A.[6] This manuscript is listed as the *Cod. 17 Bibliothecae S. Catherinae Pisarum.* Manuscript D was discovered by F. Stegmüller in 1931, which showed a similarity to A and C.[7] Manuscripts A, C and D expressly mention Siger as the author of the treatise, whereas Manuscript B makes no mention of the author. It is argued, however, that the doctrinal content and literary expression reveals such a resemblance between the four manuscripts that no one could doubt the work to be

---

1. The various Averroistic, Dominican, and Augustinian treatises on the eternity of the world are listed in G. Sajó, *Un traité récemment découvert de Boèce de Dacie De Mundi Aeternitate* (Budapest: Akadémiai Kiado, 1951), pp. 64-65, n. 70.
2. For a summary of the discovery of the works of Siger, see F. Van Steenberghen, *Les oeuvres et la doctrine de Siger de Brabant* Bruxelles: Academic Royale de Belgique, 1938), XXXIX, p. 10. See also M. Grabmann, *Neuaufgefundene Werke des Siger von Brabant und Boetius von Dacien,* Sitzungsberichte der Bayerischen Akadamie, Philos. Abteilung, 1924; *Neuaufgefundene "Quaestionen" Sigers von Brabant zu den Werken des Aristotles, Miscellanea Fr. Ehrle,* I, (Rome: 1924), I, 103-47. On the authenticity of the text, see the Introduction to the critical edition of W. J. Dwyer, *L'Opuscule de Siger de Brabant "De Aeternitate Mundi,"* (Louvain: Editions de l'Institut Supérieur de Philosophie, 1937), pp. 3-25.
3. C. Potvin, *Siger de Brabant, Bulletin de l'Academie Royale des Sciences, ...* XLV, (Belgique: 1878), 330-57.
4. P. Mandonnet, *Siger de Brabant et l'averroisme latin au XIIIe siècle,* (1 éd., Fribourg: 1899), p. cli, n.5; pp. 118-20.
5. P. Mandonnet, *Siger de Brabant et l'averroisme latin au XIIIe siècle,* II éd., *Les Philosophes Belges,* VI-VII, Louvain: Editions de l'Institut Supérieur de Philosophie, 1908-1911), pp. iii-iv; 131-142. Mandonnet indicates that P. Glorieux considered Manuscript B to be a work of Godfrey of Fontaine. See P. Glorieux, *Un recueil scolaire de Godefroid de Fontaines.* (Paris Nat. Latin, 16297), *Recherches de théologie ancienne et médiévale,* III, 1931, pp. 37-53.
6. F. Pelster, *Die Bibliothek von Santa Caterina zu Pisa, eine Büchersammlung aus den Zeiten des hl. Thomas von Aquin, Xenia Thomistica III,* (Rome: 1925), III, 249-80.
7. F. Stegmüller, *Neugefundene Quaestionen Siger von Brabant, Recherches de théologie ancienne et médiévale,* III, 1931, pp. 158-82.

Siger's.[8]

Though the doctrine in the four manuscripts is the same, Manuscript B is considered to be superior because of its briefness, order, and style (revealing Siger in his definitive writing). Manuscripts A, C, and D, on the other hand, contain vague formulas, superfluous expressions, prolonged development; in addition, all those lack unity. These manuscripts are believed to be characteristic of an oral teaching or the report of a student. A further point of difference between the manuscripts might be noted. Manuscript B is divided into three parts which form the three sections of the treatise; Manuscripts A, C, and D are divided into four parts.[9]

The date of the writing of Siger's treatise is not universally agreed upon by scholars. Mandonnet prefers the year 1270;[10] Glorieux, who had attributed the work to Godfrey of Fontaine, prefers the year 1272.[11] Pelster thinks that it could not have preceded the year 1271;[12] and Dwyer's conclusion is that it was written either at the end of 1271 or the end of 1272.[13] On the basis of this evidence, St. Thomas Aquinas' treatise, the *De Aeternitate Mundi,* was written before Siger's; there is general agreement among scholars that St. Thomas' treatise was written between the years 1270 and 1271, during the Averroist controversy at the University of Paris.[14]

The *De Aeternitate Mundi* of Siger of Brabant is written in the style and literary structure of a *Quaestio,* much like the *Opuscula* treatises of St. Thomas, and somewhat like the *Quaestiones Disputatae.* The

8. Dwyer, *op. cit.,* pp. 13-19.
9. There is no definite resolution of the relation of the four manuscripts, Dwyer, *op. cit.,* p. 12. On the other hand, Van Steenberghen offers a reserved conclusion: Manuscript A is the work of Siger; Manuscript B is a report of a student. See *Siger de Brabant d'après ses oeuvres inédites. I. Les oueuvres inédites, Les Philosophes Belges,* XII, (Louvain: Editions de l'Institut Supérieur de Philosophie, 1931), p. 14, n. 3.
10. P. Mandonnet, *op. cit.,* I, p. 114.
11. P. Glorieux, *op. cit.,* pp. 37-53.
12. F. Pelster, *Die Uebersetzungen der Aristotelischen Metaphysik in den Werken des hl. Thomas von Aquin, Gregorianum,* XVII, 1936, pp. 381 ff. Pelster notes that St. Thomas cites Book A as Book XI (1270-1271), after which date he calls it Book XII. This change was due to a new Latin translation of the *Metaphysics* of Aristotle by William of Moerbeke. Siger adopted this change in the *De Aeternitate Mundi,* and therefore it could not have been written before the year 1271.
13. Dwyer, *op. cit.,* p. 8.
14. Mandonnet dates the treatise 1270; Grabmann and Van Steenberghen date the treatise 1271. See E. Gilson, *Le Thomisme* (Paris: Vrin, 1945), p. 533. The condemnation of 1270 listed 13 theses attributed to Aristotle and his commentators, one of which concerned the eternity of the world. See *Chartularium Universitatis Parisiensis,* I, n. 432. The condemnation of 1270 was not effective, and Averroism was still taught at the University of Paris. In 1277, the "great" condemnation listed 219 theses taken from Aristotle, Averroes, Siger of Brabant, Boethius of Dacia, St. Thomas Aquinas, Giles of Rome, Bacon, and certain Neoplatonists. See *Chartularium Universitatis Parisiensis,* I, n. 556.

authority of Aristotle is dominant in the treatise, and Averroes is also cited. The treatise is, therefore, primarily an elaboration of some points of doctrine taken chiefly from Aristotle's *Metaphysics*. The main problems handled in the treatise may also be found in other works of Siger, especially in his commentaries on the *Metaphysics* and *Physics* of Aristotle.[15] The *De Aeternitate Mundi* of St. Thomas, on the other hand, has as its prime purpose to establish the possibility of an eternal creation. In his treatise, St. Thomas makes no refutation of Aristotle's arguments for an eternal world as upheld by Averroes and Siger of Brabant.[16] St. Thomas is more concerned with disproving the Augustinian theologians at the University of Paris who professed a temporal creation.[17] These Augustinian theologians are the "murmurers," of whom St. Thomas is extremely critical when he says that they speak as though wisdom originated in their own minds.[18] St. Thomas was convinced that these Augustinians were teaching Platonic doctrine in the name of St. Augustine.[19]

There are at present four different published editions of the treatise: the two editions of Mandonnet already mentioned; an edition of

---

15. See F. Van Steenberghen, *Les oeuvres et la doctrine de Siger de Brabant,* pp. 82-83, for a chronology of the works of Siger. The chief works of Siger may be found in the following editions. P. Mandonnet, *Siger de Brabant et l'averroisme de XIIIe siecle,* 2nd ed. This edition contains the *Quaestiones de Anima Intellectiva; Tractatus de Necessitate et Contingentia Causarum; Quaestio utrum haec sit vera: homo est animal nullo homine existente; Quaestiones Naturales.* F. Van Steenberghen, *Siger de Brabant d'après ses oeuvres inédites. I. Les oeuvres inédites.* This edition contains the *In de Generatione et Corruptione; In Libros Tres de Anima.* C. Baeumker has edited the *Impossibilia* in *Die "Impossibilia" des Siger von Brabant. Eine philosophische Streitschrift aus dem XIII Jahrhundert, Beiträge,* II, 6, (Münster: 1898). Siger's commentary on Aristotle's *Physics* may be found in P. Delhaye, *Siger de Brabant, questions sur la physique d'Aristote, Les Philosophes Belges,* XV, (Louvain: Editions de l'Institute Supérieur de Philosophie, 1941.) The commentary on Aristotle's *Metaphysics* may be found in C. A. Graiff, *Siger de Brabant, questions sur la métaphysique, Philosophes médiévaux,* I, (Louvain: Editions, de l'Institut Supérieur de Philosophie, 1948).
16. St. Thomas' treatise is not a direct attack on Siger's position since St. Thomas wrote his treatise before Siger. Nor does Siger make mention of St. Thomas' arguments for the possibility of an eternal creation. Siger's treatise is predominantly a reaffirmation of Aristotelian principles, and more especially the necessity of eternal generation. St. Thomas' criticism of the necessity of eternal generation may be found in *Sum. Theol.,* I, 46, 1-2; *Contra Gentiles,* II, 31-38; *De Potentia,* III, 17; *Quodlibet.,* III, q.14, a.2; *In Metaph.,* XII, lect. 5; *In II Sent.,* d.1, q.1, a.5.
17. For example, see St. Bonaventure, *In II Sent.,* d.1, pars 1, a.1, q.2, concl., *Opera Omnia,* Quaracchi, 1902, Vol. II, p. 22; *In Hexaëm,* VI, 4, Quaracchi, Vol. V, p. 361; *Breviloquium,* II, 1, 3, *Tria Opuscula,* Quaracchi, 1911, p. 61.
18. St. Thomas Aquinas, *De Aeternitate Mundi,* in *S. Thomase Aquinatis Opuscula Philosophica;* ed. Marietti, n. 307, p. 108.
19. On the Franciscan and Dominican controversy and the literature written in the 13th century, see P. Glorieux, *Le Correctorium Corruptorii "Quare," Bibliotheque Thomiste,* Vol. IX, (Belgique: Kain, 1927).

P. Barsotti;[20] and the edition of W. J. Dwyer.[21] In his edition of the treatise, Dwyer used Manuscript B in a corrected version, indicating the errors of the copyists, and supplying at the bottom of the page the variations from Manuscripts A, C, and D. The critical edition of Dwyer was used for the English translation of Siger's *De Aeternitate Mundi.*

The basic question of Siger of Brabant in the *De Aeternitate Mundi,* is whether the human species or any species can begin to be. Siger proposes two arguments which would lead one to conclude that every species must have a beginning. The first of these arguments is based on the fact that no species has existence except in singulars. If, therefore, any individual of some species has been created when it had not existed before, the species must be created when it had not existed before. The second argument is based on the fact that universals have no existence except in singulars, and since God created singulars, He also created universals which must therefore have had a beginning.

In his refutation of these two arguments, Siger proposes three questions which constitute the three divisions of the *De Aeternitate Mundi.* (1) How is the human species caused? (2) How do universals exist in singulars? (3) Does potency precede act, or does act precede potency in duration?

The first question dealing with the way in which the human species is caused begins with the statement that the human species, according to philosophers, is made by God through generation.[22] Siger then distinguishes between essential and accidental generation.[23] Essential generation proceeds from determined and individual matter, through the transmutation of a thing from non-being to being, or from privation to form. Species cannot be caused through an essential generation because determined matter is not applied to species.[24] Though not generated essentially, the human species is generated

20. P. Barsotti, *Siger de Brabantia De Aeternitate Mundi,* Series scholastica, XIII, (Münster: Aschendorff, 1933). This edition reproduces Manuscript A with variations and corrections taken from Manuscripts B, C and D; and for the first time the four manuscripts of the treatise are found in a critical edition.

21. Dwyer, *op. cit.*

22. There is no production outside of generation. See Aristotle, *De Gener. et corrup.,* I, 3, 317b, 16; II, 9, 335a 25; *De Caelo,* I, 10, 279b 12; *Metaph.,* III (B), 4, 999b 5; VII (Z), 8, 1033b 5, IX, 8, 1049b 24; Averroes, *In Metaph.,* XII, com. 18; Venetiis apud Junctas, 1562, fol. 143; *Tahafut Al-Tahafut* (*The Incoherence of the Incoherence*), Disc. I, nn. 61-63; 101-102; tr. S. Van den Bergh, (Oxford; University press, 1954), I, 35-36; 59-60. See also Siger, *In de Gen. et Corrup.,* I, 15; ed. Van Steenberghen, I, 273; I, 20, p. 274; I, 21, p. 275; II, 9, p. 287; *In Metaph.,* III, 16; ed. Graiff, pp. 145-156; *In Phys.,* III, 6; ed. Delhaye, p. 199.

23. See Averroes, *In Metaph.,* XII, com. 8, fol. 140; com. 18, fol. 143. See also Siger, *In de Gen. et Corrup.,* I, 16; Van Steenberghen, Vol. I, p. 273; *In Metaph.,* III, 15; Graiff, p. 135; III, 16, p. 150; VII 7, p. 372.

24. See Aristotle, *Metaph.,* VII (Z), 8, 1033a 23-1033b 29. See also Siger, *In Metaph.,* VII, 8; Graiff, pp. 373-74.

accidentally through the generation of any individual, and not through the generation of only one determined individual.[25] The human species is eternal and caused,[26] according to philosophers, because in the individuals of the human species one is generated before the other eternally. In this way the species can be and be caused only through an individual's existing and being caused.[27] To say that species was at one time nonexistent, would be impossible and would amount to saying that a certain individual began when there was no previous individual.[28]

The notion of time is similar to the notion of species. Aristotle says that although every second is finite, in time there is a *then* before the *then* to infinity.[29] On the basis of this reasoning, what is composed of things finite in quantity yet infinite in number has to be infinite.[30] Siger, therefore, concludes that just as past time has to be through a certain *then,* so also species has to be through the existence of any one of its individuals.

The second question of the *De Aeternitate Mundi* deals with the universal. The problem arises because of one of the two objections proposed in the beginning of the work, in which it was stated that universals have their existence in singulars. Siger argues that according to

---

25. Siger is opposing those who say that the human species is not able to have been made eternal by God unless it had been made or created in some determined or eternal individual, as the species of heaven was made eternal. Siger argues that just as man can be abstracted in thought from individual matter and from the individual, so he can be abstracted in existence. Socrates is *this* man and he is also *a* man. Aristotle's example, to which Siger appeals on this point, is that generating a brass sphere generates a sphere, because a brass sphere is a sphere, *Metaph.,* VII (Z), 8, 1033a 30-1033b 9. See Averroes, *Tahafut Al-Tahafut,* also Siger, *In Metaph.,* III, 15; Gaiff, p. 136; VII, 9, p. 374.

26. Since the human species is generated eternally, it cannot be caused in the sense that it can proceed directly from God. Siger is following Avicenna's principle that from unity only unity can proceed. See Avicenna, *Metaph.,* tr. IX, c.4, *ad fin.,* in *Metaphysica sive eius prima philosophia,* (Venice: 1495). See also Aristotle, *De Gener. et Corrup.,* II, 10, 336a 28; Siger, *De Nec. et Cont. Causarum;* ed. Mandonnet, pp. 111-12; *Impossibilia,* I; ed. Baeumker, p. 5; *In Metaph.,* V, 10-11; Graiff, pp. 301-305; III, 16, p. 142; in *Phys.,* I, 37; Delhaye, p. 75; III, 6, p. 199.

27. The perpetual generation of individuals thus insures the eternity of species. See Averroes, *In Phys.,* VIII, com. 46, fol. 176. See also Siger, *In Metaph.,* III, 13; Graiff, p. 119.

28. To ask whether it is true that man is an animal when no man exists presupposes an impossible conclusion because there are always men and humanity, *Quaestio utrum haec sit vera: homo est animal nullo homine existente;* ed. Mandonnet, p. 69. See Aristotle, *Metaph.* IV, 2, 1003b 25: man, a man, an existing man are one.

29. Aristotle, *Phys.,* IV, 11, 219b 9; 22a 5; IV, 13, 222a-222b 8; VIII, 1, 251b 10; VIII, 8, 263b 20; VIII, 10, 266a 10. See Averroes, *In Phys.,* VIII, com. 11, fol. 157. See also Siger, *In Phys.,* VIII, 10; Delhaye, pp. 206-208.

30. Aristotle, *Phys.,* III, 4, 203b 15; III, 6, 206a 25. See Siger, *In Phys.,* I, 40; Delhaye, pp. 80-81.

Aristotle, universals in themselves exist in the mind.[31] Themistius, too, says that concepts are universals which the mind collects and stores within itself;[32] and the genus is a certain concept gathered from the slight similitude of the singulars.[33] On the other hand, universals are universal things or they could not be said of particulars, and in this sense universals are not in the mind. Siger's effort in this question will be to explain the meaning of the universal as universal, and the relation of the universal to particulars.

The universal as universal, and because it is a universal, is not a substance.[34] The universal can be spoken of in two ways: something which is denominated universal, e.g., a man or a stone which exists universally in the nature of things, and which is not in the mind; and the intention of universality which is in the mind, and which is understood universally and abstractly. Man is a universal because he is known universally and abstractly from individual matter; the abstracted comprehension of things, though a common understanding of particulars, is not a thing in itself but is in the mind because it is universal. Universals, as universals, therefore, according to Averroes, have their sole existence as intelligibles and not as being,[35] and that is why Themistius says that universals are concepts.[36] Further, the universal in actuality is intelligible in actuality, and the intelligible in actuality is also the intellect in actuality.[37] The universal in actuality, need not exist before it may be known. Only the intelligible in potency precedes the understanding of it, and this is universal in potency.[38] The universal is not a universal

---

31. Aristotle, *De Anima*, II, 5, 417b 23.
32. Themistius, *Paraphrases Aristotelis, De Anima*, II, 5; ed. L. Spengel, (Leipsig: Teubner, 1866), II, 103, lines 9-11.
33. Themistius, *Paraphrases Aristotelis, De Anima*, I, 1; ed. L. Spengel, II, 6, lines 14-18.
34. Aristotle, *Metaph.*, VII, 13, 1038b 9; VII, 10, 1036a 8; III, 6, 1003a 5. Aristotle is arguing against the Platonists who place universality in things.
35. Averroes, *In de Anima*, I, 8, in *Averrois Cordubensis Commentarium Magnum in Aristotelis de Anima Libros;* ed. F. S. Crawford, (Cambridge: The Mediaeval Academy of America, 1953), pp. 11-12; *In de Anima*, III, 18, p. 440; *In Metaph.*, XII, 2, fol. 292; VII, 24, fol. 174; VII, 43, fol. 196; VIII, 6 fol. 214; *Tahafut Al-Tahafut*, Disc. I, nn. 110-11, pp. 65-66; Disc. III, n. 185, p. 111. Aristotle does not explicitly say that universals exist only in the mind, but that knowledge always refers to the universal, *Metaph.*, VII, 6, 1031b 20. See also Siger, *In de Anima*, I, 8; ed. Van Steenberghen, Vol. I, pp. 34-37; *In Phys.*, I, 7; Delhaye, pp. 27-28; II, 6, pp. 89-91; In *Metaph.*, III, 13; Graiff, pp. 117-19; III, 15, pp. 128-34; VII, 15, pp. 379-81.
36. Themistius, *Paraphrases Aristotelis, De Anima*, II, 5; Spengel, II, 103, lines 9-11.
37. Aristotle, *De Anima*, III, 4, 430a 4; a 22; III, 7, 431b 17; Averroes, *In de Anima*, III, 15; ed. Crawford, p. 434; III, 36, p. 480; Siger, *In de Anima*, III, 4; Van Steenberghen, p. 128.
38. Aristotle, *De Anima*, III, 4, 429a 12; a 22; III, 5, 430a 1; a 20; Averroes, *In de Anima*, III 5; Crawford, pp. 387-88; Siger, *In de Anima*, III, 5; Van Steenberghen, pp. 128-29.

before the concept and the act of understanding;[39] universals, as universals, are entirely in the mind and are generated neither essentially nor accidentally. The nature which is understood universally is in particular things and is generated accidentally. Though universals as universals are understood universally and abstractly, there is nevertheless a conceptual unity and persistence of universals because of their existence in particulars, and as long as there are particulars, there is a continuous potential understanding in particulars.

In the third question of the treatise, dealing with the priority of potency and act, Siger attempts to explain Aristotle's treatment of the problem.[40] In line with the reasoning of Aristotle, the priority of potency and act in the order of generation and in the order of time is resolved in the following way. Potency is prior to act in generation, because generation proceeds from potency to act and from the imperfect to the perfect. Thus, in considering something numerically the same which has existence at some time in potency and sometimes in act, potency precedes act. Act precedes potency in time, however, because before the being in potency there is another of the same species in act, educing it from potency to act. When act and potency are referred to species, both are eternal;[41] neither simply precedes the other in time, but one comes before the other to infinity, e.g., man is always in act and always able to be man.[42] Thus, according to Aristotle, the prime mover and prime matter are eternal; just as God always exists, so does the potential man since he is regarded as in prime matter.[43] Siger's conclusion is in line with the reasoning of Aristotle and Averroes. The universe of caused being must be eternal.[44] If the universe of caused being were not at some time, then potency would precede act, which is impossible. Further, if everything were in potency, and there were no being in act, all being would be in potency, and matter of itself would come into act; which is also impossible. Aristotle, therefore, had to explain the order of generation by appealing to prime movers as causes of the species of things which are educed from potency to act by them.[45]

39. Aristotle, *De Anima*, III, 5, 429a 24; Averroes, *In de Anima*, III, 5; Crawford, pp. 387-88; III, 14, p. 429; III, 18, p. 437; Siger, *In de Anima*, III, 18-19; Van Steenberghen, pp. 149-51.

40. For Aristotle's position, see *Metaph.*, IX, 8, 1049b ff. See also Siger, *In Metaph.*, III, 27; Graiff, pp. 178-81.

41. The eternity of act and species was treated by Siger in his answer to the first question. See Aristotle, *Metaph.*, VII, 8, 1033 16; IX, 8, 1050b 1. Siger, *In Metaph.*, VII, 8-9; Graiff, pp. 373-74. The eternity of potency refers to prime matter. See Aristotle, *Phys.*, 1, 5, 188a 17; 1, 9, 192a 25-33; *Metaph.*, VII, 3, 1029a 20-25.

42. See p. 80, n. 29.

43. Aristotle, *Metaph.*, IX, 8, 1049b 24.

44. Aristotle, *Phys.*, I, 8, 191b 12; Averroes, *Tahafut Al-Tahafut*, nn 99, 104-105, pp. 58, 61-62.

I would like to express my deepest gratitude to Father Gerard Smith, S.J. and Dr. John O' Riedl for their encouragement and inspiration throughout the years. I am also grateful to Dr. James H. Robb for his careful checking of the manuscript and for many helpful suggestions. To Miss Astrid Richie, formerly of Marquette University, I am indebted for her assistance in the preparation of this text.

---

45. See Aristotle, *Metaph.*, XII, 6, 1071b 3-1072a 18; Averroes, *In Metaph.*, XII, com. 54, fol. 339; Siger, *In Metaph.*, III, 27; Graiff, pp. 178-81.

Siger of Brabant

Question on the Eternity of the World

The first question is whether the human species and in general the species of all individuals began to exist only by way of the propagation of generable and corruptible things when it had no previous existence whatsoever; and it seems that this is so.

The species of which any individual began to exist when it had had no previous existence at all is new and began to have existence since it universally and entirely had had no previous existence. The human species is such, and in general the species of all individuals generable and corruptible, because every individual of this type of species began to exist when it had had no previous existence. And, therefore, any species of such things is also new and began, since in all cases it had not previously existed. The major is stated thus: because the species does not have being nor is caused except in singulars and in causing singulars. If, therefore, any individual of some species has been created when it had not existed before, the species of those beings will be such a kind.

Secondly, this same conclusion is also able to be reached in a different manner thus: universals, just as they do not have existence in singulars, so neither are they caused. Every being is caused by God. Therefore, if man has been caused by God, since he is some being of the world, it is necessary that he come to exist in a certain determined individual; just as the heaven and whatever else has been caused by God. Because, if man does not have an individual eternity, as has the sensible heaven according to philosophers, then the human species will have been caused by God so that it began to exist when it had not existed before.

To prove this one must consider, in the first place, how the human species was caused, and in general any other universals of generable and corruptible things; and in this way an answer should be made to the question and the aforementioned argument.

Secondly, since the foregoing argument admits that universals exist in singulars, one must seek or consider how this may be true.

Thirdly, because some species began to exist when it had surely not existed before, and because it follows that potentiality precedes act in duration, it should be seen which of these preceded the other in duration. For, this presents a difficulty within itself.

I

Concerning the first, therefore, we should know that the human species has not been caused, according to philosophers, except through generation. Now, because in general the being of all things is in matter which is in potency to form, they are made by a generation which is either essential or accidental. From this, however, that the human species has been made by God through generation, it follows that it does not proceed directly from Him. The human species, however, and in general the species of all things which are in matter, since it is made through generation, is not generated essentially but accidentally. It is not generated essentially, because if any one were to study those things which are made universally, then every thing which is made is made from this determined and individual matter. For, although arguments and knowledge are concerned with universals, yet operations are regarding singulars. Now, however, determined matter does not pertain to the meaning of species, and therefore is not generated essentially; and this is held by Aristotle in VII *Metaphysicae.*[1] The same reason why form is not generated is also the reason why the composite which is species is not generated. And I call the species a composite, just as Callias in his own nature is this soul in this body, so also animal is soul in body. The common nature of form and species that they are not generated essentially is because individuated matter pertains to the consideration or reasoning of neither of the things from which generation essentially comes, through the transmutation of the thing from non-being to being, or from privation to form. The human species, however, although not generated essentially has nevertheless been generated accidentally, because it thus happens if man, just as he has been abstracted in thought from individual matter and from the individual, so he might be abstracted in existence. Then, just as he is not generated essentially, it might be thought that he is also not generated accidentally; but, because man in his being is this man, Socrates or Plato, then Socrates is also a man, as Aristotle says in VII *Metaphysicae,*[2] that generating a brass sphere generates a sphere because a brass sphere is a sphere. And since just as Socrates is a man, so is Plato, and so with the others. Hence it is that man is generated through the generation of any individual, and not only of one determined individual.

Now, from the explanation it is clear in what way the human species is considered by philosophers eternal and caused. For, it is not to be thought of as eternal and caused as if it existed abstracted from individuals. Nor is it eternally caused in the sense that it exists in an eternally caused individual, as the species of heaven or an intelligence; but rather because in the individuals of the human species one is generated before

---

1. Aristotle, *Metaphysics,* VII, 8, 1033b 5.
2. Aristotle, *Metaphysics,* VII, 8, 1033a 30.

the other eternally, and the species has to be and to be caused through an individual's existing and being caused. Hence it is that the human species always exists and that it did not begin to be after previous non-existence. For, to say that it began to be after it had not existed before is to say that there began to be a certain individual before whom no other individual of that species had existed. And since the human species has not been caused otherwise than generated through the generation of individual before individual, the human species or that which is called by the name of man begins to exist because universally everything generated begins to exist - begins, nevertheless, to exist when it existed and had previously existed. For, man begins to be through the generation of Socrates who is generated; he exists, nevertheless, through the existence of Plato of the previous generation. Those things are not contradictory about the universal; just as there is nothing repugnant for a man to run and not to run. Indeed, man runs in the person of Socrates, and man does not run in the person of Plato. From the fact, nevertheless, that Socrates runs, it is not true to say that man universally and entirely does not run. So also, in the fact that Socrates is generated man begins to be, is not to say that man begins to be in such a way that the had not in any wise previously existed.

From the previous discourse the solution to the aforementioned argument is clear.

And first, it must be said that this argument as just stated, namely, that that species is new and began to exist when it had not previously existed, must be denied. Any individual of this kind began to be when it did not previously exist because even though it be true that no individual man began to be after not existing, yet no individual of this kind begins to be unless another one had previously existed. Species does not have existence so much through the existence of one of its individuals as another, and so the human species does not begin to be when it had not existed before. For, to admit that the species is such is to say that not only a certain individual of it began to be when he had not been before, but any individual of it began to be when neither he nor another individual of that species had existed before.

And the given reason is similar to the reasoning by which Aristotle speculates in IV *Physicorum*,[3] whether past time is finite. All past time whether near or remote is a certain *then,* and the certain *then* has a measured distance to the present *now;* therefore all past time is finite. And each of the aforementioned propositions is clear from the meaning of that *then* of in IV *Physicorum.*[4] The solution of this reasoning, according to Aristotle, is that although every second is finite, nevertheless

---

3.   Aristotle, *Physics,* IV, 13, 222a 10.
4.   Aristotle, *Ibid.*

since in time there is a *then* before the *then* to infinity, therefore not all past time is finite. For, what is composed of things finite in quantity yet infinite in number has to be infinite. So also, although there is no individual man but that he has begun to exist when he had not existed before, yet there is an individual before the individual to infinity; it is thus that man does not begin to be when he had in no wise existed before, and neither does time. And the case is similar - just as past time has to be through a certain *then,* so also species have to be through the existence of any one of its individuals.

Finally, as regards the form of the reasoning as proposed in the second way, it must be said that the universal does not have existence nor is caused except in singulars; since it is also said that all being has been caused by God, it must be conceded that man also exists as a being of the world and caused by God. But, since it is brought in the discussion and inferred that man has come into existence in some determined individual, it must be said that this conclusion is in no wise to be drawn from the premises; indeed, that reasoning is a hindrance to itself. For, it is accepted in the first place, that man does not have existence except in singulars nor is caused except in singulars; and it is clear that according to this reasoning he has existence and is created through one or through another. For this reason, therefore, it must be concluded that it is reasonable that man has come into existence in some determined individual. Indeed, the human species comes and came into being accidentally by the generation of individual before individual to infinity. This is not to say, however, that it (the human species) comes into existence only in some determined individual and when it had not existed before. Whence we should wonder about those arguing thus since they want to argue that the human species had begun through its being made; and yet that it was not made essentially but rather by the making of the individual, as they confess. To show their intention they ought to show that individual has not been generated before individual to infinity. This, however, they do not show but they propose one false theory, that the human species is not able to have been made eternal by God unless it had been created in some determined and eternal individual, just as the species of heaven was made eternal; and when they find no eternal being among the individuals of man, they think that they have demonstrated that the whole species began to exist when it had not been at all before.

II

The second question is whether universals are in particulars, and it is clear that they are not since Aristotle says in *II De Anima,*[5] that universals in themselves exist in the mind. And Themistius in a similar book,[6] says that concepts are similar things which are universals which the mind collects and stores within itself. And the same Themistius, in *super principium De Anima,*[7] says that genus is a certain concept gathered from the slight similitude of the singulars; the concepts however are in the conceiving mind, and universals, since they are concepts, are also in the mind.

But on the other hand, universals are universal things for otherwise they might not be said of particulars; and for this reason universals are not within the mind.

Moreover, the thing itself, which is the subject for universality, the man or the stone, is not in the mind. Also, the intention of universality must consist in its being called and denominated universal; and hence man and stone since they are called universals, the intention of universality is in these. Either both, the thing and the intention, or neither is in the mind. Because, if man and stone in respect to the fact that they are, are not in the mind, it seems that neither are they there in respect to the fact that they are universals.

The solution. The universal, because it is a universal, is not a substance, as Aristotle states in VII *Metaphysicae.*[8] And so this is clear. The universal, in that it is a universal, is different from any singular. If, therefore, the universal, in that it is universal, would be a substance, then it would be differing in substance from any of the singulars, and each (singular) would be a substance in act, both singular and universal; the act however, would be distinguished. Therefore universals would be distinct substances and separated from particulars; on this account with Aristotle it amounts to the same thing to say that universals are substances as to say that they are separated from particulars. And if the universal, in that it is universal, is not a substance, then it is evident that there are two things in the universal, namely, the thing which is denominated universal, the man or stone, which is not in the mind, and the intention itself of universality, and this is in the mind; so that the universal in that it is universal does not exist except in the mind, as is evident in this way. For, nothing is called a universal because it exists of its own nature commonly and abstractedly from particulars, or by the work of the intellect in the nature of things;

---

5. Aristotle, *De Anima,* II, 5, 417b 23.
6. Themistius, *Paraphrases Aristotelis, De Anima,* II, 5; ed. L. Spengel, (Leipsig: Teubner, 1886), II, 103, lines 9-11.
7. Themistius, *Paraphrases Aristotelis, De Anima,* I, 1; ed. L. Spengel, II, 6, lines 14-18.
8. Aristotle, *Metaphysics,* VII, 13, 1038b 9.

because if in its own nature, in its very being, it were to exist abstractly from particulars, it would not be spoken of them since it would be separated from them and we would not need an active intellect. Moreover, the active intellect does not give things any abstraction in existence from individual matter or from particulars, but gives to them an abstraction according to intellection by producing an abstract intellection of those things. If, therefore, the man or stone are universals, it is not except that these things are known universally and abstractly from individual matter. These things do not exist thus in the nature of things because if understood, those things, the man and the stone, do not have existence except in the mind. Since the abstract comprehension of these things is not in things, then those things, because they are universals, are in the mind. And this can also be seen in like cases.

A certain thing is said to be known because there is a knowledge of it and it happens that it is understood. The thing itself, however, with respect to what it is, although it be outside the mind, yet in respect to its being understood, that is, insofar as there is understanding of it, exists only in the mind. Because, if universals are universals, and they are understood as such, namely, abstract and common to particulars, then the universals as universals do not exist except in the mind. And this is what Averroes says in *super Illum De Anima,*[9] that universals as universals are entirely intelligibles; not as beings but as intelligibles. The intelligible, however, as intelligible, that is, insofar as there is an understanding of it, is entirely in the soul. Thus also Themistius says that universals are concepts.[10]

But it must be observed that the abstract and common understanding of any nature, although it be something common, as a common understanding of particulars, yet is not common according to its being predicated of particulars in that it has to be abstracted from particulars; but that which is abstractly and commonly understood and of consequence is so signified, is spoken concerning particulars. For this reason: because that very nature which is spoken of and comprehended as a general thing is in things and is therefore spoken concerning particulars. Although those things are known and understood abstractly and commonly, they do not exist as such; therefore things of this kind are not predicated of particulars according to the ideas of genus and species.

And one must also consider that it is not necessary that the universal exist in actuality before it may be known because the universal in actuality is intelligible in actuality. Now, it is one and the same actuality whether of the intelligible in actuality or of the intellect in actuality; just as it is one motion whether of the active or of the passive, although

---

9. Averroes, *Super Illum De Anima,* cap. 3, n. 18; ed. Venetiis apud Junctas, 1562, fol. 16.
10. Themistius, *Paraphrases Aristotelis, De Anima,* II, 5; ed. L. Spengel, II, 103, lines 9-11.

they be different. But the intelligible in potency certainly precedes the understanding of it; however, such a thing is not universal also except in potency, and so it is not necessary that the universal have to be universal except in potency before it is understood.

Nevertheless, some have held the contrary in this discussion because the very activity of understanding precedes in the natural order the object causing that act. Now, however, the universal, in that it is universal, moves the intellect and is the object which causes the act of understanding; on this account it seems to them that the universal is not universal in that it is so understood, indeed, because the universal in the natural order is universal before it is so understood and is the cause of that understanding of it.

But the solution of this is that that nature by which is caused the act of the intelligible and of the intellect, which is the intellect in act, is the active intellect and also the phantasm which naturally precede that act. In what manner, however, those two concur to cause the act of understanding must be sought in *super Illum De Anima.*[11] But this must be said that the universal is not a universal before the concept and the act of understanding, as at least that act is of the active intellect. For, the understanding of the thing which is in the possible intellect, since it is possible as regards the subject, belongs to the active intellect as efficient. Thus the universal does not have formally that which is universal from the nature which causes the act of understanding. Indeed, as has been mentioned before, it is the concept and the actuality from which the universal receives its universality. Therefore universals, in that they are universals, are entirely in the mind. On this account they are not generated by nature inasmuch as they are universals, neither essentially nor accidentally. For the nature which is stated and understood universally is in particular things and is generated accidentally.

To the first objection it must be said that the fact that universals are universal things can be understood in two ways; either because they exist universally or because they are understood universally. Universals, however, are not universal things in the first manner as if they existed universally in the nature of things, for they then would not be concepts of the mind. But universals are universal things in a second manner, that is, they are understood universally and abstractly; in this way universals, insofar as they are universals and since they are concepts, cannot be spoken of particulars as such. For, the idea of genus or species is not said of them, but the very nature which is thus understood as that which is itself included, is not in the mind, and is said of particulars.

---

11. Siger of Brabant, *In III De Anima,* III, 14; ed. Van Steenberghen, p. 172.

In regard to another point, it must be said that things are rightly named after something which does not exist in reality. For, a thing understood is named from the understanding of it which is not in it but in the mind; and so also the universal is named from the universal and abstract understanding of it which is not in it but in the mind.

III

Consequently we must investigate the third question. Although act precedes potency in thought, for potentiality is defined through act, as we say the builder is able to build, potency nevertheless is prior to act in substance and in perfection in a thing which proceeds from potency to act because the things which are later in generation are, in substance and perfection, prior, since generation proceeds from the imperfect to the perfect and from potentiality to act. Act is also before potentiality in substance and perfection in the respect that potentiality and act are looked at in different ways; because eternal things are prior to corruptible things in substance and perfection. But nothing eternal, in respect that it is such a thing, is in potentiality. In corruptible things, however, there is an admixture of potency.

The question is whether act precedes potentiality in time, or potentiality the act.

And it seems that the act does not precede potentiality in time because in eternal beings one is not before the other in time. But when the act of a certain species and the potentiality to that act are looked upon according to the species, they are both eternal. For, man is always in act and is always able to be man. Therefore the act thought of in relation to the species does not precede potentiality in time.

Moreover, in this matter in which one is to come from the other in a cycle to infinity, there is none which is first in time. But the seed is from the man and the man from the seed to infinity. Therefore, in those things the one does not precede the other in time. Just as in the case of the seed from which a man is generated there is another generating man previously existing, so also previous to that generating man, since he himself was generated, there must have been a seed from which he was generated.

What is first in the order of generation is first in the order of time. But potentiality is prior to act in the order of generation since generation proceeds from potentiality to act, and therefore it is prior in the order of time.

Moreover, there is no reason why act should precede potentiality in time except that by a power a being is made in act through some agent of its own kind existing in act. But, although from this it follows that the act of the agent precedes in time the act and perfection of the generated thing by that agent, nevertheless, it does not seem to happen that the act of the one generating precedes in time that which is in potentiality to the act of generation. Nor from this also does act simply precede potentiality in time, although some act precedes some potentiality to that act. For, just as being in potentiality comes into actuality through something of its own species in act, so also the thing existing in act in that species is generated from something existing in potentiality to the act of that species. For, just as that which is in potentiality, namely a man, is brought into act by a man in act, so also the man generating is generated from the previous seed and from a man in potency; and so in that reasoning the hen has preceded the egg in time and the egg the hen, as people argue.

On the other side is Aristotle in IX *Metaphysicae.*[12] For, he holds that although what proceeds from potentiality to act is the same in number, yet potentiality precedes the act in time, nevertheless, the same being in relation to species and existing in act precedes potentiality.

Moreover, everything existing in potentiality is brought into actuality through something existing in act and at length is brought into the order of moving things by a mover existing completely in act who did not previously have in his power to be anything except in act. Therefore, according to this, act is seen simply to precede potency in time.

To prove this we must first consider that something numerically the same which has existence at some time in potentiality and some time in act is able to be prior in time than it is. But because this potency is preceded by act in another, since every being in potentiality comes into actuality by that which is in some way of its species, therefore it is not proper to say simply that potentiality precedes act in time.

Secondly, one must consider that if the whole universe of caused beings were at some time not being, as certain poets, theologians and natural philosophers claimed, Aristotle says in XII *Metaphysicae,*[13] then potentiality would precede act simply. And also if some entire species of being, as the human species, would begin to exist when it had never existed before, just as some think they have demonstrated, the potentiality for the actuality of that species would simply precede the act. But each of these is impossible, as is evident from the first consideration.

For, if the whole universe of beings at some time had been in potentiality, so that none of the beings would be totally in act - always an agent in act and the mover - then the beings and the world would not now be except in potentiality, and matter of itself would come into act, which is impossible. Thus Aristotle says in XII *Metaphysicae,*[14] and so does his

---

12. Aristotle, *Metaphysics,* IX, 8, 1049b 5.
13. Aristotle, *Metaphysics,* XII, 6, 1071b 24.
14. Aristotle, *Metaphysics,* XII, 6, 1072a.

Commentator,[15] that for things to be at rest in an infinite time and afterward to be in motion is the same as for matter to be self-moving.

From the second question it is evident that this is impossible. For, since the prime mover and agent is always in act, and something in potency is not prior to something in act, it follows that it always moves and acts and makes anything or does anything without an intermediate movement. From this, however, that it is always moving and so acting, it follows that no species or being proceeds to actuality, but that it has proceeded before, so that the same species which were, return in a cycle; and so also opinions and laws and religions and all other things so that the lower circle around from the circling of the higher, although because of the antiquity there is no memory of the cycle of these. We say these things as the opinion of the Philosopher, although not asserting them as true. One, nevertheless, should notice that a certain species of being is able to go into act when it did not exist except in potentiality, although at another time it also was in act, as is evident. For, it happens in the heavens that a certain spectacle and constellation appear in the heavens previously not existing, the effect of which is properly another species of being here below, which is then caused and which yet previously existed.

Thirdly, it must be considered that when it is taken that the potency to an act and the act educing that potency are of the same kind in the generator and thing generated, it is not said in so taking them that act precedes potentiality simply nor potentiality act, unless the act is taken according to the species and the proper potentiality is taken according to the individual. For, a man in act, and a certain man in act, inasmuch as he is generating, precedes in time that which is being in potency, namely, man generated. But because in this order, just as being in potency proceeds into act through something existing in act, and so act precedes any given potentiality, so also everything existing in act in this species goes from potentiality to act, and so potentiality in this species precedes any given act. Therefore neither simply precedes the other in time, but one comes before the other to infinity, as was stated.

In the fourth question we must consider that in a certain order of moving and acting beings it is necessary that that thing which proceeds from potentiality to act comes to some act that educes that potentiality to actuality, and this act does not have to go from potentiality to act. Therefore, since every being in potentiality goes to actuality through some being of its own species in act, not all being, however, in actuality and generating proceeds from potentiality to act. Hence it is that in any given being in potentiality to some act, the act of the species in a certain way, although not entirely for the same reason, precedes that potentiality in

15. Averroes, *Commentaria in Librum XI (12) Metaph. Arist.,* cap. 1, n. 29; ed. Venetiis apud Junctas, 1562, fol. 313.

time; not however in any given being in act does potentiality from which it
proceeds to act, precede. And, therefore, the act is simply said to precede
potentiality in time, as has been explained, namely, because the first
mover leading into act all being in potentiality does not precede in time
the being in potentiality, since the being in potentiality is regarded in the
rank of prime matter. For, just as God always exists, according to
Aristotle, so also does the potential man, since he is regarded as in prime
matter. Moreover, the prime mover does not precede in time the being in
potentiality, since it is looked upon as in matter properly considered in
relation to species, as man is in the seed. For, it is never true, according
to Aristotle, to say that God existed, unless potential man existed or had
existed, as in the seed. But in a third manner from what has been said,
act simply precedes potentiality in time because in any being in
potentiality, as given in proper matter, the act of that potentiality having
to educe the potency to act, precedes in time. It is not thus with any given
being in act that the potentiality to that act precedes it in time, as is
evident in prime movers educing to actuality all beings in potentiality. In
the aforementioned we utilize, as also does Aristotle, prime movers as
species of things which are educed from potentiality to actuality by them;
and unless they were the beings of a certain kind in act which do not
proceed from potentiality to actuality, the act would not simply
precede the potentiality in time, as Aristotle has said in IX
Metaphysicae,[16] saying that act precedes potentiality in time, adding the
reason, because one act is always taken as before another up to the one
which is always the prime mover.

From this the solution of the reasoning of those opposed is clear.

To the first problem, therefore, it must be said that being in
potentiality is not eternal unless when it is regarded as in prime matter.
For, when taken as in its proper matter, according to which anything is
said to exist properly in potentiality, as is said in IX Metaphysicae,[17] it is
new, unless it were taken according to species. For, just as nothing generated is corruptible in infinite time, so also nothing generable is not generated in infinite time since the generable has been taken as in proper
matter and in a position near to generation, as the Commentator says in
super Ium Caeli et Mundi.[18]

To the second problem it must be said, as had been mentioned,
that in the order of things generating existing in act which also proceed
from potentiality to act, there is no being in act before the being in
potentiality, but one there is always before the other to infinity. Because

---

16. Aristotle, Metaphysics, IX, 8, 1050b3.
17. Aristotle, Metaphysics, IX, 8, 1049a 7.
18. Averroes, Commentaria in Lib. I De Caelo et Mundo, cap. II, pars quarta, n. 121; ed.
    Venetiis apud Junctas, 1562, fol. 82.

every being in potentiality in the essential order of moving and acting beings at length comes to some being existing in act which does not go from potentiality to act; hence it is that on account of that order the act is said simply to precede that potency.

To the third problem it must be said that it is well established that in a being which is the same in number proceeding from potency to act, potency precedes act; but that, nevertheless, before the being in potency there is another of the same species in act, educing it from potency to act.

To the last problem we must say that it is truly spoken that act precedes potentiality, because all being in potentiality goes into actuality through something existing in act. Nor do those two things which are contradictory hinder one another. In the first place, this is not so because the being in act educing that which is in potentiality into act precedes in time not only the act in the being generated, but also the potentiality proper to the actuality of the being generated because of the fact that not only is the act of the generated being from the one generating, but also the being in potentiality to the act of the generated being is also from the one generating, as the seed from the man. And universally, proper matters are from the prime mover educing each thing from potentiality to act. In the second place, what is opposed does not hinder, as is evident from what has been said above. Although in the order of moving beings, on the basis of which the argument is made, it is necessary to admit that before being in act, there is a being in potency from which it proceeds into act, so also before being in potency there is a being in act which educes itself from potentiality to act; nevertheless, in another order of moving things it is necessary to hold that there is a being in act which educes into act what is in potency, since the being in potency from which it is made does not precede it, as is evident.

BIBLIOGRAPHY

Aristotle. *Aristoteles graece ex recensione I. Bekkeri,* ed. Academia Regis Borussica. 2 vols. Berlin: Reimer 1831.

_____. *The Works of Aristotle;* tr. W. D. Ross and others. 11 vols. Oxford, Clarendon 1928-1931.

_____. *Aristotle's Metaphysics, a Revised Text with Introduction and Commentary,* by W. D. Ross. Oxford: Clarendon 1924.

_____. *Aristotle's Physics, a Revised Text with Introduction and Commentary,* by W. D. Ross. Oxford: Clarendon 1936.

Averroes. *Aristotelis Stagiritae Libri Omnes . . . cum Averrois Cordubensis Variis in Eosdem Commentariis.* Venetiis apud Juntas, 1562-1574.

_____. *Averrois Cordubensis Commentarium Magnum in Aristotelis de Anima Libros,* ed. F. S. Crawford. Cambridge: The Mediaeval Academy of America, 1953.

_____. *Tahafut Al-Tahafut (The Incoherence of the Incoherence),* tr. S. Van den Bergh. Oxford: University Press 1954.

Avicenna. *Metaphysica Avicennae sive eius prima philosophia.* Venice: 1495.

Baeumker, C., *Die "Impossibilia" des Siger von Brabant Eine philosophische Streitschrift aus dem XIII Jahrhundert, Beiträge,* II, 6. Münster: 1898.

Barsotti, P. *Siger de Brabantia De Aeternitate Mundi,* ("Series scholastica," XIII.) Münster: Aschendorff, 1933.

Bonaventure, St. *Commentaria in Quatuor Libros Sententiarum Petri Lombardi,* in *Opera Omnia,* Vol. II. Quaracchi: Ex Typographia Collegii S. Bonaventura 1882-1902.

_____. *In Hexaëmeron,* in *Opera Omnia.* vol. V.

_____. *Breviloquium,* in *Tria Opuscula. Quaracchi: Ex Typographia* Collegii S. Bonaventurae 1911.

Burbach, M. "Early Dominican and Franciscan Legislation Regarding St. Thomas," in *Mediaeval Studies,* IV (1942), 139-58.

Chartularium Universitatis Parisiensis, ed. Denifle-Chatelain. 4 vols. Paris: 1889-1897.

Chenu, M. D. *Introduction à l'étude de saint Thomas d'Aquin.* Montréal: Institut l'Etude Médiévaux, XI, 1950.

Chevalier, J. *La notion du nécessaire chez Aristote et chez ses prédécesseurs.* Paris: Alcan, 1915.

Chollet, A. "Aristotelisme de la scolastique," *Dictionnaire de théologie catholique,* I, coll. 1869-1887.

Chossat, M. "St. Thomas d'Aquin et Siger de Brabant," *Revue de philosophie,* XXIV, 553-75; XV, 25-53.

Delhaye, P. *Siger de Brabant, questions sur la physique d'Aristote, Les Philosophes Belges,* XV. Louvain: Editions de l'Institut Supérieur de Philosophie, 1941.

Duhem, P. *Le système du monde.* Vol. V. Paris: Librare scientifique A. Hermann et Fils, 1917.

Dwyer, W. J. *L'Opuscule de Siger de Brabant "De Aeternitate Mundi."* Louvain: Editions de l'Institut Supérieur de Philosophie, 1937.

_____. "Le texte authentique du 'De Aeternitate Mundi' de Siger de Brabant," in *Revue Néoscolastique de philosophie,* XL, (1937), 44-66.

Forest, A. *La structure métaphysique du concret selan s. Thomas d'Aquin.* Paris: Vrin, 1931.

Gilson, E. *Le Thomisme.* Paris: Vrin, 1945.

_____. *The Christian Philosophy of St. Thomas Aquinas.* New York: Random House, 1956.

_____. *Dante et la philosophie.* Paris: Vrin, 1939.

_____. *Being and Some Philosophers.* Toronto: Pontifical Institute of Mediaeval Studies, 1949.

_____. *History of Christian Philosophy in the Middle Ages.* New York: Random House, 1954.

Glorieux, P. Le Correctorium Corruptorii "Quare," in *Bibliotheque Thomiste.* Belgique: Kain 1927.

_____. *Un recueil scolaire de Godefroid de Fontaines.* (*Paris Nat. Latin,* 16297), *Recherches de théologie ancienne et médiévale,* III, 1931.

Goichon, A. M. *La distinction de l'essence et de l'existence d'après Ibn Sina.* Paris: Desclée de Brouwer, 1937.

Grabmann, M. *Neuaufgefundene Werke des Siger von Brabant und Boetius von Dacien, Sitzungsberichte der Bayerischen Akademie der Wissenschaften,* Philos. Klasse, 1924, 2.

_____. *Neuaufgefundene "Quaestionen" Sigers von Brabant zu den Werken des Aristoteles, Miscellanea Fr. Ehrle.* I. Rome: 1924, Vol. I.

_____. *Forschungen über die lateinischen Aristoteles übersetzungen des XIII Jahrhunderts, Beiträge,* XVII, 5-6, Münster: 1916.

Graiff, C. A. *Siger de Brabant, questions sur la métaphysique, Philosophes Médiévaux.* I. Louvain: Editions de l'Institut Supérieur de Philosophie, 1948.

Jolivet, R. *Essai sur les rapports entre la pénsee grecque et la pénsee chrétienne.* Paris: Vrin, 1931.

Lottin, O. "Liberté humaine et motion divine. De saint Thomas d'Aquin à la condamnation de 1277," in *Recherches de théologie ancienne et médiévale,* VII (1935), 52-69; 156-73.

Mandonnet, P. *Siger de Brabant et l'averroisme latin au XIIIe siècle,* 1 ed. Fribourg: 1899.

_____. *Siger de Brabant et l'averroisme latin au XIIIe siècle,* 2 ed. *Les Philosophes Belges.* VI-VII. Louvain: Editions de l'Institut Supérieur de Philosophie 1908-1911.

Maurer, A. *"Esse* and *Essentia* in the Metaphysics of Siger of Brabant," in *Mediaeval Studies,* VIII (1946), 68-86.

_____. "John of Jandun and the Divine Causality," in *Mediaeval Studies,* XVII (1955), 185-207.

Owens, J. *The Doctrine of Being in the Aristotelian Metaphysics.* Toronto: Pontifical Institute of Mediaeval Studies, 1951.

Paulus, J. *Henri de Grand. Essai sur les tendances de sa métaphysique, Etude de Philosophie Médiévale.* XXV. Paris: Vrin, 1938.

Pelster, F. *Die Bibliothek von Santa Caterina zu Pisa, eine Büchersammlung aus den Zeiten des hl. Thomas von Aquin. Xenia Thomistica.* III. Rome: 1925.

_____. *Die Uebersetzungen der Aristotelischen Metaphysik in den Werken des hl. Thomas von Aquin, Gregorianum,* XVII, 1936.

Potvin, C. *Siger de Brabant,* (Bulletin de l'Academie Royale des sciences,...) XLV, Belgique: 1878.

Salman, D. "Notes sur le première influence d'Averroes," In *Revue Néoscolastique de philosophie,* XL (1937), 1-40.

Sajó, G. *Un traité récemment découvert de Boèce de Dacie De Mundi Aeternitate.* Budapest: Akadémiai Kiado, 1954.

Siger of Brabant. *Quaestiones de Anima Intellectiva,* in P. Mandonnet, *Siger de Brabant et l'averroisme latin au XIIIe siècle.* 2 ed. *Les Philosophes Belges.* VII. Louvain: Editions de l'Institute Supérieur de Philosophie, 1908.

_____. *Tractatus De Necessitate et Contingentia Causarum;* ed. Mandonnet.

_____. *Quaestio utrum haec sit vera: homo est animal nullo homine existente;* ed. Mandonnet.

_____. *Quaestiones Naturales;* ed. Mandonnet.

_____. In *de Generatione et Corruptione,* in F. Van Steenberghen,*Siger de Brabant d'après ses oeuvres inédites. I. Les oeuvres inédites, Les Philosophes Belges.* XII. Louvain: Editions de l'Institut Supérieur de Philosophie, 1931.

_____. *In Libros Tres de Anima;* ed. Van Steenberghen.

_____. *Impossibilia,* in C. Baeumker, *Die "Impossibilia" des Siger von Brabant, Eine philosophische Streitschrift aus dem XIII Jahrundert, Beiträge.* II, 6, Münster: 1898.

_____. *In Phys.,* in P. Delhaye, *Siger de Brabant, questions sur la physique d'Aristote, Les Philosophes Belges.* XV. Louvain: Editions de l'Institut Superiéur de Philosophie, 1941.

_____. *In Metaph.*, in C. A. Graiff, *Siger de Brabant, questions sur la métaphysique, Philosophes Médiévaux,* I, Louvain, 1948.

_____. *De Aeternitate Mundi,* in W. J. Dwyer, *L'Opuscule de Siger de Brabant "De Aeternitate Mundi."* Louvain: Editions de l'Institut Supérieur de Philosophie, 1937.

Stegmuller, F. *Neugefundene Quaestionen Siger von Brabant, Recherches de théologie ancienne et médiévale, III,* 1931.

Themistius. *Paraphrases Aristotelis, De Anima;* ed. L. Spengel. Leipsig: Teubner, 1866.

Thomas Aquinas, St. *Summa Theologiae.* 5 vols. Ottawa: Impensis Studii Generalis O. Pr., 1941-1945.

_____. *Summa Contra Gentiles.* Rome: Marietti, 1934.

_____. *Scriptum super Libros Sententiarum Magistri Petri Lombardi;* ed. R. P. Mandonnet. 3 vols. Paris: Lethielleux, 1929-1933.

_____. *In Metaphysicam Aristotelis Commentaria,* ed. M. R. Cathala and R. M. Spiazzi. Turin: Marietti, 1950.

_____. *In Libros Physicorum,* in *Opera Omnia,* ed. Leonine. Vol. II.

_____. *Quaestiones Disputatae.* 2 vols. Turn: Marietti, 1949.

_____. *Quaestiones Quodlibetales.* Turin: Marietti, 1927.

_____. *S. Thomae Aquinatis Opuscula Omnia necnon Opera Minora.* Vol. I, *Opuscula Philosophica,* ed. J. Perrier, Paris: Lethiellex, 1949.

_____. *S. Thomae Aquinatis Opuscula Philosophica.* Turin: Marietti, 1954.

Van Steenberghen, F. *Les oeuvres et la doctrine de Siger de Brabant,* Bruxelles: Acad. Royale de Belgique, XXXIX, 3, 1938.

_____. *Siger de Brabant d'après ses oeuvres inédites. I. Les oeuvres inédites, Les Philosophes Belges.* XII. Louvain: Editions de l'Institut Supérieur de Philosophie, 1931.

Vaux de, R. *Notes et textes sur l'Avicennisme latin.* Paris: Vrin, 1934.

_____. "La première entree d'Averroes chez les latins," in *Revue des Sciences Philosophique et Théologique,* XXII (1933), 3-20.

St. Bonaventure

Selected Texts

*On the Eternity
of the World*

Translated from the Latin
With an Introduction

By

PAUL M. BYRNE, L.S.M., Ph.D.
Assistant Professor of Philosophy
Marquette University

Translator's Introduction

John Fidanza, who was to become St. Bonaventure, was born in 1221 in the small town of Bagnorea, near Viterbo in Italy.[1] The exact date of his entry into the relatively new religious order founded by St. Francis of Assisi, the Order of Friars Minor, is uncertain. It is known that he was quite young, and the year 1238, at the age of seventeen, is most probable. It is known, however, that he was a student at Paris prior to 1245 since he studied under Alexander of Hales who died in that year.

In 1248, St. Bonaventure began his own public teaching in theology. He had been since 1245 a *baccalaureus,* teaching under the direction of a master. The secular masters in the University of Paris were attempting at that time to exclude the new orders, the Franciscans and Dominicans. Thus, although Pope Alexander IV on Oct. 23, 1256 had ordered the University to receive both St. Bonaventure and St. Thomas Aquinas as Doctors within the University, they were not accepted as such until a year later, in October, 1257. By this time St. Bonaventure's teaching career had ended and he had entered on a new one as Minister General of His Order. For the remainder of his life, he devoted himself chiefly to the work of administration. In 1273 he was appointed Bishop of Albano and made a Cardinal of the Church. He died July 15, 1274 while attending the Council of Lyons.

The chronology of his principal works may be conveniently divided into two periods: an early one, from 1250 to 1259, covering his teaching career; and a later period, from 1267 to his death, a time of great controversy within the University of Paris.[2] His earliest work composed about 1250-1255, is the *Commentary on the Sentences* of Peter Lombard. The *Breviloquium* dates from about 1255-1257. In 1257 he published the *De Mysterio Trinitatis* and the *De Scientia Christi.* One of his best known works, the *Itinerarium Mentis in Deum* was composed in 1259. During the later period, the period of controversy arising out of the spread of Averroism within the Faculty of Arts as well as out of the "innovations" of St. Thomas Aquinas, St. Bonaventure gave a series of conferences or meditations at Paris. These included the *Collationes de Decem Praeceptis* in 1267-1268, the *De Decem Donis Spiritus Sancti,* and most important of all for his position on the eternity of the world, the *Collationes in Hexaemeron* in 1273, a year before his death.

---

1. For the life and thought of St. Bonaventure, see E. Gilson, *La philosophie de S. Bonaventure* (3rd ed., Paris: J. Vrin, 1953); English translation: *The Philosophy of St. Bonaventure* (New York: Sheed and Ward, 1938).
2. For the dating of the works listed, see Fr. Jules D'Albi, O.F.M. Cap., *S. Bonaventure et les luttes doctrinales de 1267-1277* (Tamines: Duculot, 1922), pp. 60-62, pp. 145 ff.

These two periods in the works of St. Bonaventure coincide with two different phases of the story of Aristotle in the thirteenth century. Some twenty years before St. Bonaventure began to teach, the influence of Aristotle's philosophy within the University of Paris had begun to grow. In spite of various prohibitions and regulations in 1210, 1215, and 1231, the Stagirite's work in physics and metaphysics permeated everywhere and did not cease to gain ground, along with the works of the Arabian commentators.[3] By 1250 when St. Bonaventure began his *Commentary on the Sentences,* all the works were well known to him and used by him. It has been estimated that in that *Commentary,* there are over one thousand references to the Aristotelian works, covering the whole range of his writings.[4]

It was, of course, also well known to the Franciscan theologian that Aristotle, in teaching the eternity of motion, taught the eternity of the world. Accordingly, as he is commenting on the second book of the *Sentences,* he asks the question: Has the world been produced in time or from eternity?[5] In this early period it is merely a matter of another error to be refuted, and the principal opponent is Aristotle. Seventeen years later, in 1267, the situation is quite different. The Averroistic interpretation of Aristotle had so taken hold in the Faculty of Arts that it was no longer a question of the errors of an ancient pagan philosopher, but of the very possibility of a Christian philosophy, as Bonaventure understood it, that was at stake. In that year, 1267, St. Bonaventure pointed out, in his *Collationes in Decem Praeceptis,* the three philosophical errors at the heart of the Averroist controversy: that the world is eternal, that there is one intellect for all men, and that it is impossible for a mortal to attain immortality.[6] In that same work he refers to the change that had occurred since his youth: "When I was a student I heard it said that Aristotle posited the world as eternal, and when I heard the reasons and arguments to this effect, my heart began to pound and I began to think: how can this be? But now this is all so public that no hesitation (in the matter) is permitted."[7] The theological reaction to this version of Aristotle had now set in, a reaction that was to lead to the condemnations of 1270 and 1277. At least one of his historians, Jules D'Albi, believes that the Franciscan theologian was the initiator of an anti-Averroist, and to

---

3. See E. Gilson, *History of Christian Philosophy in the West* (New York: Random House, 1955), p. 245.
4. *Ibid.,* p. 686, n. 18.
5. St. Bonaventure, *In II Sent.* d.1, p.1, a.1, q.2; *Opera omnia* (Quarrachi: Collegii S. Bonaventurae 1882), vol. II, p. 21. All subsequent references to works of St. Bonaventure are to this edition.
6. St. Bonaventure, *Collationes in Decem Praeceptis II,* 25, vol. V, p. 514.
7. *Ibid.,* II, 28, vol. V, p. 515.

some extent anti-Thomist, movement.[8]

While the works of Bonaventure's later period often refer in an incidental way to the error of an eternal world, the *Collationes in Hexaemeron* contain the most detailed and strongest attack on that error. Consequently, in considering his views on the eternity of the world, it is necessary to look first at the discussion in the Commentary on the Sentences and then note the attitude adopted in the *Collationes in Hexaemeron.*

The position Bonaventure attacks in the *Sentences* is first stated in terms of the Aristotelian doctrine that everything that begins to be begins to be by motion or change. Since the absolutely perfect, and so circular, motion of the heavens could not have preceded itself or its *primum mobile,* it must be eternal. Moreover, if motion comes to be, it comes to be through motion. In order to avoid an infinite regress, a motion lacking a beginning, and so eternal, must be posited. The next statement of the Aristotelian position is in terms of time. In every time there is a before and after, and an instant or moment only through that before and after. Consequently, if time began, it began in time and there would follow the absurdity that there was time before time began. Turning now from the world to God, two more arguments for the eternity of the world are given. Given an adequate and actual cause, the effect is given. But God from eternity is the adequate and actual cause of this world. Consequently, the world must be from eternity. Further, if God were to begin to produce the world, He would pass from rest to act and thus be subject to mutability. But to say this is to deny His absolute perfection and simplicity.

Bonaventure will not, as St. Thomas does, concede that a beginning in time is indemonstrable. Not only does he think that his refutations of the Aristotelian arguments are valid, he holds that it is contradictory to admit that a created universe may have existed from all eternity. Thus, he argues that one cannot add to the infinite; but the new day, the new month, the solar and lunar revolutions would do just that if the world's duration were infinite. Moreover, there could be no order in that duration, one revolution before another, if there were not a first revolution. Besides this, the infinite cannot be traversed; but the world, on Aristotle's supposition, would have traversed an infinite number of revolutions. Further, since the world would not be without there being men, either there would be in an infinite time an actually infinite number of rational souls (and there can be no actually infinite number), or there has been a transmigration of souls (which Aristotle would not admit), or there is one soul for all men (an even more grievous error). Finally,

---

8. D'Albi, *op. cit.,* pp. 139ff.

Bonaventure gives the argument which he seems to regard as the fundamental one. The world, produced by God out of nothing as an origin, has
being after nonbeing, and so cannot be eternal.

From this argument it is evident that he believes that the type of
production which is creation involves a temporal product. For him, the
creating of things, which he calls a supra-natural motion, involves the
concreating of both natural motion and time. Since Aristotle knew only
of natural motion, he did not know this. Furthermore, creation is from
the divine will through the divine wisdom, which knows that eternity does
not befit the mutable nature of the creature. Thus, Bonaventure believes
that *rationes necessariae,* arguments necessitating assent, may be given for
what the Christian already knows by faith and revelation: *In the beginning
God made all things.*[9]

Several times during his conferences on the six days of creation,
the *Collationes in Hexameron,* the Minister General of the Franciscans
refers to the error of an eternal world. In the sixth conference, Aristotle
is represented to be the leader of a group (presumably the Latin
Averroists) who are likened to the angels of darkness.[10] The Franciscan
theologian attempts to identify the first step these men have taken on the
path to darkness. He finds it in Aristotle's rejection of the Platonic Ideas,
the exemplars of things in their First Cause. From this root error spring
all the others: God knows only Himself, has no knowledge of the particular, moves only as desired and loved, has no foreknowledge or providence.
This leads to a fatalism which would remove all reason for penalty or
glory in an after life. Exemplarism, divine providence, and the disposition
of earthly things according to penalty or glory having been hidden away in
darkness, these men suffer from a concomitant triple blindness: about
the eternity of the world, the unicity of the intellect, and future happiness
or punishment.

The next reference to this triple blindness of Aristotle and his
followers is within a context presenting the doctrine of a divine
illumination of the intellect.[11] This time Bonaventure finds an excuse for
Aristotle, if not for the Parisian Averroists: "On the eternity of the world,
he could be excused because he understood this as a philosopher,
speaking as a naturalist (*physicus*), saying that it could not begin to be
through nature."[12] The Franciscan would agree that as a naturalist this is
just what Aristotle would have to say. But it is precisely this limitation in

---

9. St. Thomas Aquinas, *Sum. Theol.* I, 46, 2c: " *rationes non necessariae,* . . . which give rise
    to ridicule among unbelievers etc."
10. St. Bonaventure, *Collationes in Hexaemeron* VI, 2.
11. *Ibid.,* VII, 1-2.
12. *Loc. cit.*

Aristotle's thought that leads him and his followers into all the errors consequent upon the denial of the Ideas.

This "excuse" is relevant to Bonaventure's conception of the relation of Christian wisdom to Aristotle's thought. One immediately thinks of that famous passage wherein the Franciscan teacher says that Aristotle knew how to speak the language of science, but Plato spoke the language of wisdom in affirming the existence of the eternal reasons of Ideas, "while both languages, namely of wisdom and science, have been given through the Holy Spirit to Augustine."[13] His Franciscan spirituality coinciding with his intellectual formation in Augustinian wisdom gave to St. Bonaventure's whole thought a Christocentric bearing.[14] Even as St. Augustine could, in his *City of God,* express his regret that the Platonists, who were so wise, could fall into so many errors because they knew nothing of the wisdom of Christ, so also St. Bonaventure finds Aristotle in error on the origin of creatures and the destiny of man for this same reason. But he seems to regard the case of Aristotle as far more desperate than that of Plato, for Aristotle philosophizes as a "naturalist," a *physicus,* who knows nothing of the transcendent eternal reasons, the Divine Ideas, on which all wisdom must be grounded. Little wonder, then, that the Averroists in the Faculty of Arts who were willing to forget they were Christians while they were philosophizing according to Aristotle were being led into a darkness of errors, and among them, that the world is eternal.

The Translation

The translations that follow: *In II Sent.* d.1, p.1, a.1, q.2, as well as passages from the *Breviloquium* and the *Collationes in Hexaemeron,* are based on the text edited by the Franciscan Fathers of Quarrachi. The appropriate passages in the edition by Fr. Delorme[15] of the other redaction of the *Collationes in Hexameron* have been compared with the text of the *Opera omnia* edition. It seemed, with regard to the eternity of the world, no useful purpose would be served by including them, since they would be, for the most part, repetitious.

---

13. *Sermo IV de Rebus Theologicis* 18-19, vol. V, p. 572.
14. Geo. Tavard, A. A., *Transiency and Permanence, the Nature of Theology according to St. Bonaventure* (St. Bonaventure, N.Y.: Franciscan Institute, 1954), pp. 21-22.
15. *S. Bonaventurae Collationes in Hexaemeron,* ed. Ferdinandus Delorme, O.F.M. (Quarrachi: Collegi S. Bonaventurae 1934).

St. Bonaventure

In II Sent. d.1, p.1, a.1, q.2:

The question is: Has the world been produced in time or from eternity. That it has not been produced in time is shown:

1. By two arguments based on motion,[1] the first of which is demonstrative in the following way: *Before every motion and change, there is the motion of the first moveable thing (primm mobile)* ; but everything which begins to be begins by way of motion or change; therefore that motion (viz., of the first moveable thing) is before all that which begins to be. But that motion could not have preceded itself or its movable thing (*mobile*); therefore it could not possibly have a beginning. The first proposition is a basic one and its proof is as follows: It is a basic principle in philosophy[2] that "in every kind the complete is prior to the incomplete of that kind"; but movement toward place is the more perfect among all the kinds of motion inasmuch as it is the motion of a complete being, and circular motion is both the swifter and the more perfect among all the kinds of local motion; but the motion of the heaven is of this kind, therefore most perfect, therefore the first. Therefore it is evident that, etc.

2. This is likewise shown by an absurdity consequent upon the alternative. *Everything which comes to be comes to be through motion or change;* consequently, if motion comes to be it comes to be through motion or change, and with regard to this latter motion the question is similarly raised. Therefore, either there is to be an infinite regress or a positing of some motion lacking a beginning; if the motion, then also the movable thing and, consequently, also the world.

3. Similarly, a demonstrative argument based on time is as follows:[3] *Everything which begins to be either begins to be in an instant or in time.* If, therefore, the world begins to be, it does so either in an instant or in time. But before every time there is time, and time is before every instant. Consequently there is time before all those things which have begun to be. But it could not have been before the world and motion; therefore the world has not had a beginning. The first proposition is *per se* known. The second, namely that before every time there is time, is evident from the fact that if it is flowing, it was of necessity flowing

1. Aristotle, *Phys.* VIII, 1, 250b11-252b7; V, 1, 224a20-225b10. *De Gener. et Corrup.* I, 3, 317a-32-319b5.
2. Aristotle, *Phys.* VIII, 9, 265a-12-266a10. *De Caelo* II, 4, 286b10-287b22.
3. Aristotle, *Phys.* VIII, 1, 251b19-27.

beforehand.[4] Similarly, it is evident that there is time before every instant since time is a circular measure suited to the motion and the movable thing; but every point in a circle is a beginning even as it is an end; therefore every instant of time is a beginning of the future even as it is a terminus of the past. Accordingly, before every "now" there has been a past. It is evident, therefore, etc.

4. Again, this is shown by the absurdity consequent upon the alternative. If time is produced, it is produced either in time or in an instant; therefore in time. But in every time there is a prior and a posterior, both a past and a future. Consequently, if time has been produced in time, there has been time before every time, and this is impossible. Therefore, etc.

These are Aristotle's arguments based on the character of the world itself.

5. Besides these, there are other arguments based on the character of the producing cause. In general, these can be reduced to two, the first of which is demonstrative and the second based on the absurdity consequent upon the alternative. The first is as follows: *Given an adequate and actual cause, the effect is given;*[5] but God from eternity has been the adequate and actual cause of this world; therefore, etc. The major premise is *per se* known. The minor, namely that God is the adequate cause, is evident. Since He needs nothing extrinsic for the creating of the world, but only the power, wisdom and goodness which have been most perfect in God from eternity, evidently He has, from eternity, been the adequate cause. That He has also been the actual cause is evident as follows: God is pure act and is His own act of willing, as Aristotle says;[6] and our philosophers (*Sancti*) say that He is His own acting. It follows therefore, etc.

6. Also, by the absurdity of the alternative. *Everything which begins to act or produce, when it was not producing beforehand, passes from rest into act.*[7] If therefore, God begins to produce the world, He passes from rest into act; but all such things are subject to rest and change or mutability. Therefore God is subject to rest and mutability. This, however, contradicts His absolute goodness and absolute simplicity, and, consequently, is impossible. It is to blaspheme God; and to say that the world has had a beginning amounts to the same thing.

---

4. Aristotle, *Phys.* IV, 11, 220a25; IV, 12, 221a1; IV, 14, 223b12-29; IV, 13, 222a10-12.
5. Aristotle, *Phys.* II, 3, 195b17-18.
6. Aristotle, *Metaph.* XII, 5, 1071b18-19; XII, 7, 1072a21-26.
7. Aristotle, *Phys.* VIII, 1, 251a17-27; Metaph. XII, 6, 1071b7-22. St. Augustine, *Confess.* XI, 10, 12, PL 32, 814; *De Civ. Dei* XII, 17, 1, PL 41, 366.

These are arguments which the commentators and more recent men (*moderniores*) have added over and beyond the arguments of Aristotle; or, at least, they are reducible to these.

But there are arguments to the contrary, based on *per se* known propositions of reason and philosophy.

1. The first of these is: *It is impossible to add to the infinite.*[8] This is *per se* evident because everything which receives an addition becomes more; "but nothing is more than infinite." If the world lacks a beginning, however, it has had an infinite duration, and consequently there can be no addition to its duration. But this is certainly false because every day a revolution is added to a revolution; therefore, etc. If you were to say that it is infinite in past time and yet is actually finite with respect to the present, which now is, and, accordingly, that it is in this respect, in which it is finite, that the "more" is to be found, it is pointed out to you that, to the contrary, it is in the past that the "more" is to be found.[9] This is an infallible truth: If the world is eternal, then the revolutions of the sun in its orbit are infinite in number. Again, there have necessarily been twelve revolutions of the moon for every one of the sun. Therefore the moon has revolved more times than the sun, and the sun an infinite number of times. Accordingly, that which exceeds the infinite as infinite is discovered. But this is impossible; therefore, etc.

2. The second proposition is: *It is impossible for the infinite in number to be ordered.* For every order flows from a principle toward a mean. Therefore, if there is no first, there is no order; but if the duration of the world or the revolutions of the heaven are infinite, they do not have a first;[10] therefore they do not have an order, and one is not before another. But since this is false, it follows that they have a first. If you say that it is necessary to posit a limit (*statum*) to an ordered series only in the case of things ordered in a causal relation, because among causes there is necessarily a limit,[11] I ask why not in other cases. Moreover, you do not escape in this way. For there has never been a revolution of the heaven without there being a generation of animal from animal. But an animal is certainly related causally to the animal from which it is generated. If, therefore, according to Aristotle and reason it is necessary to posit a limit among those things ordered in a causal relation, then in the generation of animals it is necessary to posit a first animal. And the world has not existed without animals, therefore, etc.

---

8. Aristotle, *De Caelo* I, 12, 283a9-10.
9. St. Thomas Aquinas, *In II Sent.* d.1, q.1, a.5, ad 3m and ad 4m.
10. Aristotle, *Phys.* VIII, 5, 256a17-19.
11. Aristotle, *ibid.* ; *Metaph,* II, 2, 994a1-19.

3. The third proposition is: *It is impossible to traverse what is infinite.*[12] But if the world had no beginning, there has been an infinite number of revolutions; therefore it was impossible for it to have traversed them; therefore impossible for it to have come down to the present. if you say that they (i.e., numerically infinite revolutions) have not been traversed because there has been no first one,[13] or that they well could be traversed in an infinite time, you do not escape in this way. For I shall ask you if any revolution has infinitely preceded today's revolution or none. If none, then all are finitely distant from this present one. Consequently, they are all together finite in number and so have a beginning. If some one is infinitely distant, then I ask whether the revolution immediately following it is infinitely distant. If not, then neither is the former (infinitely) distant since there is a finite distance between the two of them. But if it (i.e., the one immediately following) is infinitely distant, then I ask in a similar way about the third, the fourth, and so on to infinity. Therefore, one is no more distant than another from this present one, one is not before another, and so they are all simultaneous.

4. The fourth proposition is: *It is impossible for the infinite to be grasped by a finite power.* But if the world had no beginning, then the infinite is grasped by a finite power; therefore, etc. The proof of the major is *per se* evident. The minor is shown as follows. I suppose that God alone is with a power actually infinite and that all other things have limitation. Also I suppose that there has never been a motion of the heaven without there being a created spiritual substance who would either cause or, at least, know it. Further, I also suppose that a spiritual substance forgets nothing. If, therefore, there has been, at the same time as the heaven, any spiritual substance with finite power, there has been no revolution of the heaven which he would not know and which would have been forgotten. Therefore, he is actually knowing all of them and they have been infinite in number. Accordingly, a spiritual substance with finite power is grasping simultaneously an infinite number of things. If you assert that this is not unsuitable because all the revolutions, being of the same species and in every way alike, are known by a single likeness, there is the objection that not only would he have known the rotations, but also their effects as well, and these various and diverse effects are infinite in number. It is clear, therefore, etc.

5. The fifth proposition is: *It is impossible that there be simultaneously an infinite number of things.*[14] But if the world is eternal and without a beginning, then there has been an infinite number of men,

---

12. Aristotle, *Metaph.* XI, 10, 1066a35.
13. St. Thomas Aquinas, *Sum. Theol.* I, q.46, a.2, ad 6m; *In II Sent.* d.1, q.1, a.5, ad 3m and ad 4m.
14. Aristotle, *Phys.* III, 5, 204a20-25. *Metaph.* 10, 1066b11

since it would not be without there being men - for all things are in a certain way for the sake of man[15] and a man lasts only for a limited length of time. But there have been as many rational souls as there have been men, and so an infinite number of souls. But, since they are incorruptible forms, there are as many souls as there have been; therefore an infinite number of souls exist. If this leads you to say that there has been a transmigration of souls or that there is but the one soul for all men, the first is an error in philosophy, because, as Aristotle holds, "appropriate act is in its own matter." Therefore, the soul, having been the perfection of one, cannot be the perfection of another, even according to Aristotle. The second position is even more erroneous, since much less is it true that there is but the one soul for all.

6. The last argument to this effect is: *It is impossible for that which has been after non-being to have eternal being,* because this implies a contradiction. But the world has being after non-being. Therefore it is impossible that it be eternal. That it has being after non-being is proven as follows: everything whose having of being is totally from another is produced by the latter out of nothing; but the world has its being totally from God; therefore the world is out of nothing. But not out of nothing as a matter (*materialiter*); therefore out of nothing as an origin (*originaliter*). It is evident that everything which is totally produced by something differing in essence has being out of nothing. For what is totally produced is produced in its matter and form. But matter does not have that out of which it would be produced because it is not out of God (*ex Deo*). Clearly, then, it is out of nothing. The minor, viz, that the world is totally produced by God, is evident from the discussion of another question.[16]

Conclusion

Whether positing that all things have been produced out of nothing would imply saying that the world is eternal or has been produced eternally.

I answer: It has to be said that to maintain the the world is eternal or eternally produced by claiming that all things have been produced out of nothing is entirely against truth and reason, as the list of the above arguments proves; and it is so against reason that I do not believe that any philosopher, however slight his understanding, has maintained this. For such a position involves an evident contradiction. But, with the eternity of

---

15. Aristotle, *Phys. II,* 2, 194a34-35.
16. St. Bonaventure, *In II Sent.* d.1, p.1, a.1, q.1.

matter presupposed, to maintain an eternal world seems reasonable and understandable, and this by way of two analogies which can be drawn. For the procession of earthly things from God is after the fashion of an imprint (*vestigium*). Accordingly, if a foot and the dust in which its print were formed were eternal, nothing would prevent our understanding the the footprint is co-eternal with the foot and, nevertheless, it still would be an imprint from the foot.[17] If matter, or the potential principle, were in this fashion coeternal with the maker, what would keep that imprint from being eternal? Rather, on the contrary, it would seem quite fitting that it should be.

Again, another reasonable analogy offers itself. For, from God the creature proceeds as a shadow, the Son as brightness. But as soon as there is light, there is immediately brightness, and immediately shadow if there should be an opaque object in its way. If, therefore, matter, as opaque, is coeternal with the maker, just as it is reasonable to posit the Son, the brightness of the Father, to be coeternal, so it seems reasonable that creatures or the world, shadow in relation to the Highest Light, is eternal. Moreover, this view is more reasonable than its contrary, viz., that matter has been eternally incomplete, without form or the divine influence, as certain philosophers have maintained. In fact, it is more reasonable to such an extent that even that outstanding philosopher, Aristotle, has fallen into this error, according to the charges of our philosophers (*Sancti*), the exposition of the commentators, and the apparent meaning of his text.

On the other hand, modern scholars say that the Philosopher has never thought this nor did he intend to prove that the world had no beginning *in any way at all,* but rather that it did not begin *by way of a natural motion.*

Which of these interpretations is the truer one I do not know. This one thing I do know, that if he held that the world has not begun *according to nature,* he maintained what is true, and his arguments based on motion and time are conclusive. But if he thought that it has *in no way begun,* he has clearly erred, as has been shown above by many arguments. Moreover, in order to avoid self-contradiction, he had to maintain either that the world has not been made, or that it has not been made out of nothing. In order to avoid an actual infinity, however, he had to hold for either the corruption of the rational soul, or its unicity, or its transmigration; thus, in any case, he had to destroy its beatitude. So it is that this error has both a bad beginning and the worst of endings.

---

17. St. Augustine, *De Civ. Dei* X, 31, PL 41, 311-12.

1. To the first objection, regarding motion, viz., that there is a first among all motions and changes because there is a most perfect one, it must be granted as true with respect to *natural* motions and changes and there is nothing against it. But with respect to the *supernatural* change, through which that *mobile* (i.e., the heaven) has proceeded into being, it is not true. For this latter change precedes every created thing, and so precedes the *primum mobile* and, of course, its motion.

2. To the objection, every motion passes into being through motion, it must be answered that motion does not pass into being *through itself* nor *in itself*, but *with another* and *in another*. And since God has, at the same instant, made the *mobile* and, as mover, acted upon the *mobile* then He has cocreated the motion with the *mobile*. If, however, you were to seek further with regard to that creating, it must be said that there a limit is reached as in the very first of things. Further on, this will be made better known.[18]

3. To the third objection, regarding the "now" of time, it must be said that just as there is a two-fold indicating of a point in a circle, either when the circle is being made or after it has been made, and just as when it is being made there is a placing and indicating of a first point but when it already is there is no locating of a first, so also with regard to time there is a twofold acceptance of the "now." In the very production of time, there has been a first "now" before which there has been no other and which was the beginning of time in which all things are said to have been produced. But with respect to time after it has been made, it is true that it is the terminus of the past and it is in the fashion of a circle. But things have not been produced in this way in a time already complete. Thus it is clear that the Philosopher's arguments do not at all establish this conclusion. With regard to the statement that before every time is time, this is true in terms of dividing time from within, but not in the sense of preceding as outside time.

4. To the objection concerning time, as to when has it begun, it must be said that it began in its own beginning (*principium*). But the source (*principium*) of time is the instant or "now"; and so it began in an instant. Also, the argument: Time has not been in an instant and so has not begun in an instant does not stand because things which are successive are not in their own beginning. This same thing can be said in another way since there are two ways of speaking about time, either according to its essence or according to its being (*esse*). If one speaks of it according to its essence, then the "now" is the whole essence of time, and that has begun to be with the mobile thing, not in another "now" but in its own self since it has been established in the very beginning and thus

---

18. St. Bonaventure, *In II Sent.* d.1, p.1, a.3, q.2, ad 5m. St. Bonaventure's point seems to be that to ask whether there is a creating of creating is to ask a meaningless question.

it has not had another measure. If one speaks of it according to its being, then it has begun with the motion of change, viz, it has not begun by way of creation, but rather through the change of the changeable, and especially of the *primum mobile.*

5. To the objection based on the adequacy and actuality of the cause, it must be said that the adequate cause of any effect is of two kinds, either as operative through its *nature* or as operative through *will and reason.* If operative through *nature,* then, as soon as it is, it produces its effect. On the other hand, if operative through *will,* even though it be adequate, it is not necessarily operative as soon as it is, since a cause of this sort is operative by way of wisdom and discernment and so takes suitability into account. Therefore, inasmuch as eternity did not befit the nature of the creature itself, it was not fitting that God should grant this most excellent of states to any thing. Accordingly, the divine will, which operates by way of wisdom, has produced the world not from eternity but in time, since, as He has produced so has He disposed and so has He willed. For He has willed from eternity to produce then when He has produced, just as I will now to hear Mass tomorrow. It is thus evident that the adequacy of the cause is not pertinent to the question.

Similarly with regard to actuality, it must be said that a cause can be in act in two ways, either in itself, as if I were to say: The sun shines, or in its effect (*in effectu*), as if I were to say: The sun illumines. In the first way God has always been in act, since He is pure act unadulterated by the merely possible. In the other way He is not always in act, for He has not always been producing.

6. To the sixth objection: If from being non-producing He has become producing, he has changed from rest to act, it must be said that there is a type of agent in which action and production add something over and beyond the agent and the producer. Such an agent is changed in some way when from non-acting it becomes acting, and in this case rest precedes operation and by operation completion is achieved. But there is another agent which is its own action, and nothing at all is added to such an agent when it produces nor does anything come to be in it which it was not beforehand. An agent of this sort neither receives completion in operating nor is it idle in nonoperating, nor, when from non-producing it comes to be producing, is it changed from rest to act. But such is God, even according to the philosophers, who have asserted god to be the most simple of beings. It is evident, therefore, that their argument is a foolish one. For if He had produced things from eternity in order to avoid idleness, then He would not be, *without things,* the perfect good, nor consequently would He *with things,* since that which is most perfect is perfect by its own self. Moreover, if because of His immutability it were necessary that things be from eternity, then He could produce nothing new now. And what kind of God would He be who is essentially incapable of any-

thing? All these consequences imply nonsense rather than philosophy or an argumentation.

If you were to ask how this is to be understood, viz., that God acts by His own self and yet does not begin to act, the answer would have to be that, even though this cannot be fully understood because of the conjoined imagination, nevertheless it can be established by argumentation necessitating assent; and anyone who withdraws from sense experience in order to consider the intelligibles will, to some extent perceive it. For if anyone were to ask whether an angel could make a pottery cup or throw a stone in spite of no hands, the answer would be that he could because he is capable, by his own power alone and without an organ, of what the soul is capable of with the body and its members. If, therefore, an angel, because of its simplicity and perfection, exceeds man to such an extent that he can do, without an organ as a means, that for which man necessarily requires an organ, can even do through one [power] what man is capable of only through many, how much more can God, who is at the very limit of the whole of simplicity and perfection,[19] produce all things without any means by the command of His own will which is nothing other than Himself, and thereby remain immutable in producing! In this way a man can be led to an understanding of this truth. But that man will grasp it more perfectly who can consider these two aspects of his Maker, namely that He is most perfect and most simple. Because He is most perfect, all the perfections there are are to be attributed to Him; because most simple, these introduce no diversity into Him, and accordingly no change or mutability. So it is, "remaining at rest He causes all things to be moved."[20]

Collationes in Hexaemeron IV, 13

The sixth division is into cause and caused; and here there are many errors. For some say that the world has been from eternity.[21] Wise men agree that something could not come to be from nothing and in this way be from eternity, since it is necessarily the case that, just as when a thing passes into nothing it ceases to be, so also when it comes to be from nothing it begins to be. But some seem to have posited an unoriginated matter, from this it follows that God does not make any-

---

19. *Liber de Causis* props. 20 and 21, ed. O. Bardenhewer, (Freiburg im Breisgau: Herder, 1882), pp. 182-83.
20. Boethius, *De Consolatione Philosophiae* III, metr. 9, ed. A. Forti Scuto, (London: Burns, Oates and Washbourne, Ltd., 1925), p. 81.
21. Aristotle, *Phys.* I, 4, 187a29.

thing. For He does not make matter, since it is unoriginated. Nor does He make form since either it comes to be from something or from nothing; not from matter since the being (*essentia*) of form cannot come from the being (*essentia*) of matter; and not from nothing since, as they suppose, God can make nothing from nothing. But let perish the notion that the power of God has matter as its supporting foundation.

Collationes In Hexaemeron V, 29

And it is in this way that the Philosopher proceeds to prove the world eternal on the basis that circular local motion, because it is perfect, precedes every motion and change. But I answer: It must be said that the perfect is before the diminished when speaking of the simply perfect, but not when speaking of the perfect in a genus, as is the case with local motion.

Collationes in Hexaemeron VI, 2-5

2. Just as it has been said of the angels, *God separated the light from the darkness,*[22] so also it may be said of the philosophers. But what was the first step that some have taken toward darkness? It was this: although all of them have seen that the first cause is the beginning of all things as well as their end, still they have differed from one another about an intermediary. For some have denied that the exemplars of things are in that first cause. The leader of this group seems to have been Aristotle who at the beginning and at the end of his *Metaphysics,*[23] and in many other places, condemns Plato's Ideas. Accordingly, he says that God knows only Himself,[24] does not need knowledge of any other thing, and moves as desired and loved.[25] On this ground these men assert that He has no knowledge of the particular. As a consequence, Aristotle especially attacks the Ideas in his *Ethics* where he says that the highest good cannot be an Idea.[26] His arguments do not hold, and the commentator [on the *Ethics,* Eustratius] has destroyed them.

3. From this error follows a further error, namely, that God does not have foreknowledge or providence since He does not have within Himself the intelligibilities (*rationes*) of things through which He might

---

22. St. Augustine, *De Civ. Dei* XI, 19, PL 41, 332-33; XI, 33, PL 41, 346-47.
23. Aristotle, *Metaph.* I, 9, 990b10 ff.; XIII, 4, 1079a4 ff.
24. Aristotle, *Metaph* XII, 9, 1074b33-1075a11.
25. Aristotle, *Metaph.* XII, 7, 1072a24-26
26. Aristotle, *Nic. Eth.* I, 6, 1096a11-1096b5.

know them. Also they say that the only truth about the future is truth about the necessary, the truth about the contingent is not truth.[27] And from this it follows that all things come to be either by chance or by the necessity of fate. Because it is impossible to come to be by chance, the Arabs introduce the necessity of fate, namely, that the sphere-moving substances are the necessary causes of all things.[28] From this it follows that the truth that there is a disposition of earthly things according to penalties and glory is obscured. For if those substances unerringly move, nothing is affirmed regarding hell, nor that there is a devil. Nor, as it seems, has Aristotle ever asserted that there is a devil, and that there is a beatitude after this life. This threefold error, then, consists in hiding away within darkness, exemplareity, divine providence and the disposition of earthly things.

4. There follows upon these a threefold blindness or mental dullness. First, about the eternity of the world, as Aristotle seems to hold according to all the Greek doctors, such as Gregory of Nyssa, Gregory Nazianzenus, Damascene, Basil, as well as according to all the Arabic commentators, who say that Aristotle taught this and they seem to speak his words. You will never find him saying that the world had a source or a beginning; rather he argued against Plato who alone seems to have held that time had a beginning.[29] And this is in direct conflict with the light of truth.

From this follows another blindness, about the unicity of the intellect. For if an eternal world is posited, some one of these consequences follows necessarily: either there is an infinite number of souls since men would have been infinite in number, or the soul is perishable, or there is a transition from body to body, or the intellect is one in all men, the error attributed to Aristotle by the Commentator.[30]

From these two it follows that after this life there is neither happiness nor punishment.

5. These men, therefore, have fallen into errors, nor have they been "separated from the darkness"; and such errors as these are the very worst. Nor has the abysmal pit been as yet locked up.[31] This is the darkness of Egypt; for although great light might be discerned in them in knowledges prior to this, still it is entirely extinguished through the aforementioned errors. And some men, seeing that Aristotle was so out-

---

27. Aristotle, *De Interp.* I, 9, 18b1-19b4.
28. St. Thomas Aquinas, *Contra Gent.* III, 87.
29. Aristotle, *Phys.* VIII, 251b16-28. *De Caelo* I, 10, 279b32-280a12.
30. Averroes, *Commentarium Magnum in Aristotelis De Anima* III, 5, ed. F. S. Crawford, (Cambridge, Mass.: Mediaeval Academy of America, 1953), pp. 387-413; III, 17, pp. 436-37; III, 36, pp. 495-502.
31. Apoc. 11: 1 ff.

standing in other matters and in them spoke the truth, cannot believe that
he has not said what is true in these matters.

Collationes in Hexaemeron VII, 1-2

1. *God saw the light, that it was good, and He divided the light
from darkness,*[32] etc. This text has been discussed in order to explain
vision by the understanding naturally bestowed upon us. He causes us to
see this, *that it was good,* both in the order of scientific consideration and
also in that of sapiential contemplation. In the order of scientific consideration, inasmuch as He illumines as light, namely, as the truth of things,
as the truth of speech, and as the truth of conduct. In the order of
sapiential contemplation, inasmuch as He illumines through the influx
into the soul of a ray from the eternal light, in order that the soul may see
that light in its own self as in a mirror, in the separate Intelligence as in a
transmitting medium as it were (*medio quodam delativo*), in the eternal
light as in a subject-source (*in subjecto fontano*). Also it has been said
that He divided the light from darkness, because certain ones have
attacked the Ideas, and as a result a threefold understanding of truth has
been hidden in darkness, namely, the truth about the eternal art, the truth
about divine providence, and the truth about the angelic fall, which follows if the angel would not have his perfection except through motion.
From this there follows a threefold blindness, namely, about the eternity
of the world, about the unicity of the intellect, and about punishment and
glory.

2. The first one Aristotle seems to posit, as well as the last one
since he does not place happiness in an afterlife; with regard to the middle one, the Commentator says he thought this. On the eternity of the
world, he could be excused because he understood this as a philosopher,
speaking as a naturalist, saying that it could not begin to be through
nature. That Intelligences would have their perfection through motion,
this he could have said inasmuch as they are not useless, because there is
nothing useless in the very ground work of nature.[33] Likewise that he
placed happiness in this life may be because, although he thought it eternal, about that he did not interject his own view (*se*), perhaps because it
was not a part of his treatment of the question. About the unicity of the
intellect it could be said that he understood the intellect to be one by reason of the inflowing light, not by reason of itself, since it is many according to subject.

---

32. Gen. 1:4.
33. Aristotle, *De Caelo* I, 4, 271a32-34. *De Anim.* III, 9, 432b21.

Breviloquium II, 1, 1-3

1. These matters regarding the Trinity having been, in a summary fashion, grasped, some remarks now have to be made about that creature that is the world. What must be held in this matter is, briefly, as follows: The whole of the earthly contrivance has been produced in being in time (*ex tempore*) and from nothing (*de nihilo*) by one, sole, and highest first Principle whose power, though itself immeasurable, has disposed *all things in a certain weight, number, and measure.*[34]

2. From these points, understood in a general way, about the production of things, truth is gained and error rejected. By "in time" is excluded the error of those positing an eternal world. By "from nothing" is excluded the error of those positing an eternal material principle. By "by one Principle" is excluded the error of the Manichees positing a plurality of principles. By "sole and highest" is excluded the error of those positing that God has produced lower creatures through the ministry of the Intelligences. By adding "in a certain weight, number, and measure" it is pointed out that the creature is an effect of the creating Trinity under a threefold causality: *efficient,* by which in the creature is unity, mode, and measure; *exemplar,* by which there is in the creature truth, species, and number; *final,* by which there is in the creature goodness, order, and weight. And, indeed, these are found in every creature, as a vestige of the Creator, whether they are corporeal or spiritual or composed of both.

3. The basis for understanding the aforesaid is as follows: For there to be perfect order and repose among things, all things must be reduced to one principle which indeed is first[35] so that it may give repose to others and most perfect so that it may give completion to all others. The first principle through which there is repose can be one and only one. This first principle, if it produces the world, since it could not produce it out of its own self, must produce it out of nothing. And because production out of nothing posits being (*esse*) after non-being (*non-esse*) on the part of the produced and immeasurability in producing power on the part of the producing principle, and since this belongs to God alone, it is necessary that the world as a creature be produced in time by that immeasurable power acting *per se* and immediately.

---

34. Wis. 11:21.
35. Aristotle, *Metaph.* II, 1 994a1-19.

MARQUETTE UNIVERSITY PRESS
Mediaeval Philosophical Texts in Translation

#1 - Grosseteste: On Light. Trans. by Clare Riedl. $5.95.

#2 - St. Augustine: Against the Academicians. Trans. by Sr. Mary Patricia Garvey, R.S.M. $7.95.

#3 - Pico Della Mirandola: Of Being and Unity. Trans. by Victor M. Hamm. $5.95.

#4 - Francis Suarez: On the Various Kinds of Distinction. Trans. by Cyril Vollert, S.J. $7.95.

#5 - St. Thomas Aquinas: On Spiritual Creatures. Trans. by Mary C. Fitzpatrick. $7.95.

#6 - Meditations of Guigo. Trans. by John J. Jolin, S.J. $7.95.

#7 - Giles of Rome: Theorems on Existence and Essence. Trans. by Michael V. Murray, S.J. $7.95.

#8 - John of St. Thomas: Outlines of Formal Logic. Trans. by Francis C. Wade, S.J. $7.95.

#9 - Hugh of St. Victor: Soliloquy in the Earnest Money of the Soul. Trans. by Kevin Herbert. $7.95.

#10 - St. Thomas Aquinas: On Charity. Trans. by Lottie Kendzierski. $7.95.

#11 - Aristotle On Interpretation: Commentary by St. Thomas and Cajetan. Trans. by Jean T. Oesterle. $7.95.

#12 - Desiderius Erasmus of Rotterdam: On Copia of Words and Ideas. Trans. by Donald B. King and H. David Rix. $7.95.

#13 - Peter of Spain: Tractatus Syncategorematum and Selected Anonymous Treatises. Trans. by Joseph Mullally and Roland Houde. $7.95.

#14 - Cajetan: Commentary on St. Thomas Aquinas' On Being and Essence. Trans. by Lottie Kendzierski and Francis C. Wade, S.J. $14.95.

#15 - Suarez: Disputation VI, On Formal and Universal Unity. Trans. by James F. Ross. $7.95.

#16 - St. Thomas, Siger of Brabant, St. Bonaventure: On The Eternity of the World. Trans. by Cyril Vollert, S.J., Lottie Kendzierski, Paul Byrne. $7.95.

#17 - Geoffrey of Vinsauf: Instruction in the Method and Art of Speaking and Versifying. Trans. by Roger P. Parr. $7.95.

#18 - Liber De Pomo: The Apple or Aristotle's Death. Trans. by Mary Rousseau.

#19 - St. Thomas Aquinas: On the Unity of the Intellect Against the Averroists. Trans. by Beatrice H. Zedler. $7.95.

#20 - The Universal Treatise of Nicholas of Autrecourt. Trans. by Leonard A. Kennedy, C.S.B., Richard E. Arnold, S.J., & Arthur E. Millward, A.M. $7.95.

#21 - Pseudo-Dionysius Aeropagite: The Divine Names and Mystical Theology. Trans. by John D. Jones.

#22 - Matthew of Vendome: Ars Versificatoria (The Art of the Versemaker). Trans. by Roger P. Parr. $9.95.

#23 - Suarez on Individuation, Metaphysical Disputation V: Individual Unity and Its Principle. Trans. by Jorge J.E. Gracia. $24.95.

#24 - Francis Suarez: On the Essence

of the Finite Being as Such, On the Existence of That Essence and Their Distinction. Trans. by Norman J. Wells. $24.95.

#25 - The Book Of Causes (Liber De Causis). Trans. by Dennis Brand. $7.95.

#26 - Giles of Rome: Errores Philosophorum. Trans. by John O. Reidl. $7.95.

#27 - St. Thomas Aquinas: Questions on the Soul. Trans. by James H. Robb. $24.95.

<Forthcoming> - William of Auvergne: De Trinitate. Trans. by Francis C. Wade, S.J., & Roland J. Teske, S.J.

Published by: Marquette University Press, Holthusen Hall, Marquette University, Milwaukee, WI 53233. **Manuscript submissions should be sent to:** Chair, MPTT Editorial Board, Dept. of Philosophy, Coughlin Hall, Marquette University, Milwaukee, WI 53233.